Classroom Companion: Economics

The Classroom Companion series in Economics includes undergraduate and graduate textbooks alike. It welcomes fundamental textbooks aimed at introducing students to the core concepts, empirical methods, theories and tools of the field, as well as advanced textbooks written for students at the Master and Ph.D. level seeking a deeper understanding of economic theory, mathematical tools and quantitative methods.

Kjetil Bjorvatn

Microeconomics Made Simple

An ABC in 1-2-3

 Springer

Kjetil Bjorvatn
Economics
NHH Norwegian School of Economics
Bergen, Norway

ISSN 2662-2882   ISSN 2662-2890 (electronic)
Classroom Companion: Economics
ISBN 978-3-032-06353-3   ISBN 978-3-032-06354-0 (eBook)
https://doi.org/10.1007/978-3-032-06354-0

Translation from the Norwegian language edition: "Mikroøkonomi - en ABC på 1-2-3" by Kjetil Bjorvatn, © Vigmostad & Bjørke AS 2025. Published by Fagbokforlaget. All Rights Reserved.

This Springer imprint is published by the registered company Springer Nature Switzerland AG
The registered company address is: Gewerbestrasse 11, 6330 Cham, Switzerland

If disposing of this product, please recycle the paper.

To Heidi and Elisabeth

Preface

I have been teaching microeconomics at the Norwegian School of Economics for a number of years, and in 2021 I wrote a book in Norwegian for this course, with a revised version coming out this year, in 2025. My students really like it, and I thought it might deserve a bigger audience. So here is an English version. Hope you like it too.

And if you do, it's in large part thanks to the wonderful contributions from my students—they've provided invaluable feedback and inspiration throughout. I want also like to thank my father, Bjarne Bjorvatn, for contributing all the illustrations, and to my good colleagues Alexander W. Cappelen, Markus Karlman, Erik Ø. Sørensen, and Bertil Tungodden for their helpful advice and encouraging comments at various stages of the writing.

Finally, I am grateful to my Norwegian publisher Fagbokforlaget for generously letting me use all the material that we developed together for the Norwegian version of the book.

December 2025

Kjetil Bjorvatn
NHH Norwegian School of Economics
Bergen, Norway

Competing Interests The author has no competing interests to declare that are relevant to the content of this manuscript.

Introduction

There are many microeconomics textbooks on the market. Some are long, others short. Some are quite mathematical, others more verbal. Some are filled with real-world examples, while others present only the theory, almost like a collection of formulas.

But there is no microeconomics textbook that tells a story—and even includes a hint of romance (with a happy ending). Until now! The first thing you'll notice when you flip through this book is that it has a cast of characters. Here they are:

Main characters:

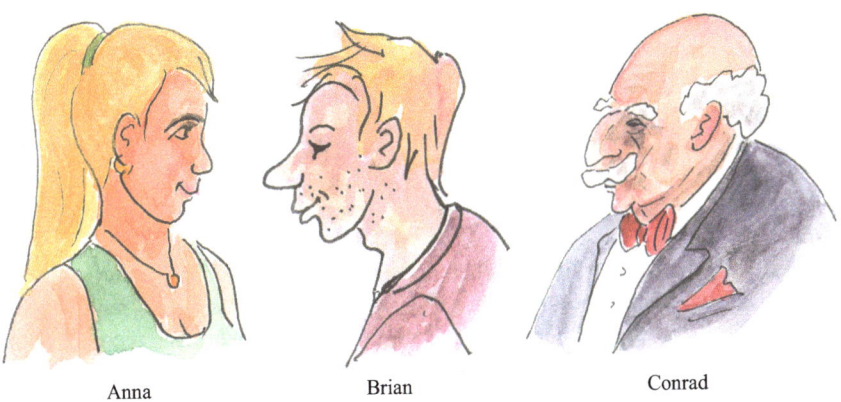

Anna   Brian   Conrad

Supporting characters:

Audrey (Brian's mother)   The professor   The competitor

Perhaps the characters (drawn by my father) remind you of someone you know. Maybe you find some of the situations funny or engaging. Take, for instance, the time Brian has more money than usual but ends up buying *less* frozen pizza than he normally does. That's an illustration of an *inferior good*, a key concept in consumer theory.

The illustrations, situations, and dialogues are meant to bring the text to life and create a sense of connection to the theory. It's easier to learn when you're having fun, and images can serve as helpful mental cues for remembering ideas. Inferior goods... let's see—yes, that was Brian and the pizza! In that way, the characters and the stories serve a clear educational purpose.

The humorous touch, however, does not mean I take the academic content lightly. This book provides a concise introduction to the core theories of microeconomics, based on standard mathematical and graphical methods. Most of the mathematics is gathered in "math boxes" to make it easier to navigate the material. Key terms are highlighted at the end of each chapter.

I've chosen not to include real-world cases or applications, even though I hope the characters and situations I've created feel realistic. The best examples are those that are current, but what's considered current depends on time and place. I'll leave it to each lecturer to spice up their teaching with their own favourite cases and examples.

The book consists of thirteen chapters divided into four parts. Part I covers consumer theory, Part II producer theory, Part III market theory, and Part IV market power and strategy. Of course, there will always be some disagreement about which topics a basic microeconomics course should include. I've chosen to focus on the material I teach in my own microeconomics course at the Norwegian School of Economics (7.5 ECTS credits), and I've tried to do that as well as I can.

Each chapter ends with four multiple-choice questions. These are checkpoints to help ensure you've understood the material before moving on. You'll find the answers at the bottom of the last page of the chapter. There is also a separate Workbook with tips and detailed solutions, closely integrated with the textbook.

Microeconomics is the foundation of all economic theory. This book, together with the Workbook, will provide you with the tools and understanding you need to analyse economic issues relevant to individuals, firms, and governments. These books also offer a solid academic foundation for more advanced and specialised courses in economics—and hopefully, inspiration to learn more.

Good luck, and enjoy!

Contents

Part I   Consumer Theory

1   Income, Prices and Preferences 3
   1.1   Introduction 3
   1.2   The Budget Line 4
   1.3   Preferences: Utility and Indifference Curves 6
   1.4   Utility Maximisation 11
   1.5   Other Types of Preferences 15
   1.6   How Does a Price Change Affect Consumption? 17
   1.7   Summary 19
   1.8   Key Terms 20
   1.9   Multiple-Choice Exercises 20

2   More About Consumer Choice 25
   2.1   Introduction 25
   2.2   Substitution Effect and Income Effect 26
   2.3   Normal and Inferior Goods 30
   2.4   Giffen Goods 32
   2.5   Fungibility 34
   2.6   Saving 36
   2.7   Summary 41
   2.8   Key Terms 42
   2.9   Multiple-Choice Exercises 42

3   Consumers at Work 45
   3.1   Introduction 45
   3.2   How Many Hours Do You Want to Work? 46
   3.3   What Job Do You Want? 50
   3.4   Wages and Labour Supply 51
   3.5   Manna from Heaven 53
   3.6   Summary 56
   3.7   Key Terms 56
   3.8   Multiple-Choice Exercises 56

4   Behavioural Economics 59
    4.1   Introduction 59
    4.2   Temptations and Saving 60
    4.3   Loss Aversion and Labour Supply 64
    4.4   Altruism and Gifts 67
    4.5   Summary 70
    4.6   Key Terms 71
    4.7   Multiple-Choice Exercises 71

Part II   Producer Theory

5   Labour and Capital 75
    5.1   Introduction 75
    5.2   Technology 76
    5.3   Isocost 78
    5.4   Production Function and Isoquant 80
    5.5   Cost Minimisation 85
    5.6   Changes in Relative Factor Prices 90
    5.7   Four Production Technologies 92
    5.8   Robots or Traditional Technology? 95
    5.9   Summary 99
    5.10 Key Terms 100
    5.11 Multiple-Choice Exercises 100

6   Costs 105
    6.1   Introduction 105
    6.2   Labour Requirements and Costs in the Short Run 107
    6.3   When the Producer Hits the Wall 112
    6.4   Cost Minimisation with Multiple Production Units 114
    6.5   Factor Requirements and Costs in the Long Run 116
    6.6   Comparison of Short-Run and Long-Run Cost Functions 118
    6.7   Summary 120
    6.8   Key Terms 120
    6.9   Multiple-Choice Exercises 121

7   Profit 123
    7.1   Introduction 123
    7.2   Profitability Defined: Profits and Operating Profits 124
    7.3   Profit Maximisation 126
    7.4   The Supply Curve 132
    7.5   Demand for Labour 134
    7.6   Summary 135
    7.7   Key Terms 136
    7.8   Multiple-Choice Exercises 136

Part III Market Theory

8 Perfect Competition 141
    8.1 Introduction 141
    8.2 Aggregating Demand 142
    8.3 The Market Demand Curve, Its Level and Slope 144
    8.4 Aggregating Supply 148
    8.5 The Market Supply Curve, Its Level and Slope 149
    8.6 Market Equilibrium 152
    8.7 Stepwise Supply 156
    8.8 International Trade 158
    8.9 Summary 160
    8.10 Key Terms 161
    8.11 Multiple-Choice Exercises 162

9 Economic Efficiency 165
    9.1 Introduction 165
    9.2 The Invisible Hand 166
    9.3 Comparative Advantage 168
    9.4 Winners and Losers from International Trade 170
    9.5 Externalities 173
    9.6 Summary 174
    9.7 Key Terms 175
    9.8 Multiple-Choice Exercises 175

10 Economic Policy 179
    10.1 Introduction 179
    10.2 Taxes and Tax Incidence 180
    10.3 Taxes and Economic Efficiency 186
    10.4 The Laffer Curve 187
    10.5 Tariffs 189
    10.6 Environmental Policy 191
    10.7 Quantity and Price Regulation 193
    10.8 Summary 195
    10.9 Key Terms 195
    10.10 Multiple-Choice Exercises 196

Part IV Market Power and Strategy

11 Monopoly 203
    11.1 Introduction 203
    11.2 Marginal Revenue 204
    11.3 Profit Maximisation 207
    11.4 How Much Market Power Does the Monopolist Really
        Have? 210
    11.5 Demand Elasticity and Monopoly Power 211
    11.6 Price Discrimination 212

11.7   Deadweight Loss from Monopoly 214
11.8   Natural Monopoly 215
11.9   Summary 216
11.10   Key Terms 217
11.11   Multiple-Choice Exercises 217

12   Oligopoly 219
12.1   Introduction 219
12.2   Three Models of Oligopoly 220
12.3   The Cournot Model 221
12.4   The Stackelberg Model 228
12.5   The Bertrand Model 233
12.6   Summary 237
12.7   Key Terms 237
12.8   Multiple-Choice Exercises 238

13   Game Theory 241
13.1   Introduction 241
13.2   Prisoner's Dilemma 242
13.3   The Stag Hunt 244
13.4   The Chicken Game 245
13.5   Sequential Games 247
13.6   A Credible Threat? 250
13.7   Summary 253
13.8   Key Terms 255
13.9   Multiple-Choice Exercises 255

Index 259

Part I

Consumer Theory

In this part of the book, we will get to know the students Anna and Brian. They are both in their first year at the School of Economics, but that's where the similarities end. While Anna has a part-time job and spends her earnings on a nice flat, exercise, and cultural activities, Brian relies on gifts from his mother and spends what little he has on beer and frozen pizza.

I use Anna and Brian to explore how consumers make choices. These choices are shaped by income, prices, and preferences. We assume that consumers maximise their utility given their budget constraint. A central question in the analysis is how changes in the choice set—that is, changes in prices or income—affect their decisions.

Chapter 1 introduces the two key tools in consumer theory: the budget line and the indifference curve. We use these to examine consumer choice based on utility maximisation.

Chapter 2 explains how a change in price affects both the relative prices of goods and purchasing power. We call these two components the substitution effect and the income effect. The direction of the first is predictable, but the second—the income effect—depends on whether the goods in question are normal or inferior. We will look at the interesting special case of inferior goods known as Giffen goods, where a price increase leads to higher demand.

Consumer theory is highly flexible, and in Chapter 3, we apply it to career choice (should you go for that demanding leadership position?) and the labour supply decision—the trade-off between leisure and consumption. How is labour supply affected by a wage increase? And how is your choice of work hours affected by unearned income, such as a grant or income from capital?

In Chapter 4, we turn to the new and exciting field of behavioural economics. Whereas standard economic theory assumes that individuals are rational (optimising based on stable preferences) and self-interested (deriving utility only from own consumption), behavioural economics challenges these assumptions by focusing on bounded rationality (we waver in our decisions and often make mistakes) and expanded preferences (we often care about others). We will use standard microeconomic tools to examine behavioural topics such as temptation, loss aversion and altruism.

Income, Prices and Preferences

1

Anna likes living comfortably, but she also wants to afford other things and is thinking about how best to spend her money.

She works part time at a grocery shop and notices that when bacon is on offer, people buy more eggs and less pancetta—and she wonders why.

1.1   Introduction

Anna is a student at the School of Economics, studying microeconomics and other exciting subjects. She is a sensible person, with a part-time job at a grocery shop and good control over her finances. Sitting in the study hall, Anna is thinking carefully about how much to spend on housing and on other things that bring her joy, such as exercise, food, and entertainment.

Anna in the study hall, looking at housing ads

© The Author(s), under exclusive license to Springer Nature Switzerland AG 2026
K. Bjorvatn, *Microeconomics Made Simple*, Classroom Companion: Economics,
https://doi.org/10.1007/978-3-032-06354-0_1

Inspired by her microeconomics course, Anna often finds herself thinking about what the people around her spend their money on—for example, what customers at the shop have in their shopping baskets. It's especially interesting when a product is on offer: How does that affect shopping behaviour?

This chapter is about consumer choice. A basic assumption in microeconomic theory is that people do the best they can with the money they have, given the prices in the market. Consumer theory is built around three key components: income, prices, and preferences.

We begin by examining the consumer's set of options, and how changes in prices and income affect these possibilities. We then turn to preferences and introduce the concept of utility, showing how the consumer can use their income to maximise utility. Finally, we consider how changes in prices and income influence consumer choice.

1.2   The Budget Line

Anna likes living well—ideally in a spacious and well-equipped flat not too far from her university. At the same time, she doesn't want to spend all her money on rent; she wants to have something left over for other things. She enjoys working out, eating well, and going to the cinema.

Anna's consumption of living space, which we shall refer to as her base (B), and everything else—which we'll group together and simply call "all other goods" (A)—is limited by her income (I). Naturally, how far her income stretches depends on the prices of the two goods, p_A and p_B. Here, we can think of p_A as the average price of everything else: for instance, if her gym membership fee goes up, p_A increases. Her budget constraint is given by:

$$I = p_A A + p_B B \quad \text{Budget constraint} \tag{1.1}$$

She illustrates the budget constraint with a diagram (Fig. 1.1). On the horizontal axis she places her home base, and on the vertical axis, all other goods. Based on current prices and her income, she draws the budget line m_1 This line represents the limit of what she can afford—her choice set.

Anna can choose any point along the m_1 line. She thinks she'll go for point b. What would you choose? Perhaps c if you place a high value on living well, or a if spacious living isn't that important to you?

Rearranging the budget constraint (1.1), we find the budget line by as follows:

$$A = \frac{I}{p_A} - \frac{p_B}{p_A}B \quad \text{The budget line}$$

The expression for the budget line shows how much of good A the consumer can buy, as a function of income, prices, and the consumption of good B.

The position of the budget line is determined by its intercepts on the two axes, which represent the maximum amount the consumer can afford of each good. The

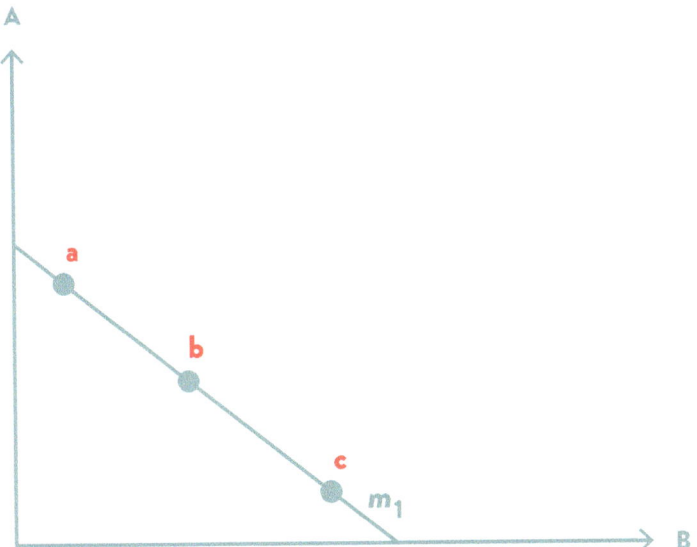

Fig. 1.1 The Budget Line. *Note* The budget line shows all possible combinations of two goods, A and B, that the consumer can afford, given the prices of the two goods and the amount of money she has available. The figure shows an example of such a budget line, which we call m_1 and three possible consumption choices: *a*, *b*, and *c*

vertical intercept is found by setting $B = 0$ in the budget line equation, which gives $A = I/p_A$—the maximum consumption of A. The horizontal intercept is found by setting $A = 0$, which gives $B = I/p_B$—the maximum consumption of B.

An increase in income shifts the budget line outward in the diagram (parallel to the original line), while a reduction in income shifts it inward, also in parallel.

The slope of the budget line is found by taking the derivative of the budget line expression with respect to B:

$$-\frac{\partial A}{\partial B} = \frac{p_B}{p_A} \quad \text{Slope of the budget line} \tag{1.2}$$

This tells us how much we must reduce consumption of good A when we increase consumption of good B, given the budget constraint. We see that the slope becomes steeper the higher the price of good B is relative to good A. If the price of good B is relatively high, this means we must give up a relatively large amount of good A in order to increase consumption of good B, within the budget available.

For example, suppose the housing market changes and it becomes more expensive to rent a flat. This means a higher price for good B and a steeper budget line, as shown by the shift from m_1 to m_2 in Fig. 1.2. Notice that the intercept on the vertical axis remains the same. Why? Because at this point Anna does not spend any money on rent, so changes in the price of housing do not affect it! However, the new maximum consumption point for the home base is, of course, lower. Of

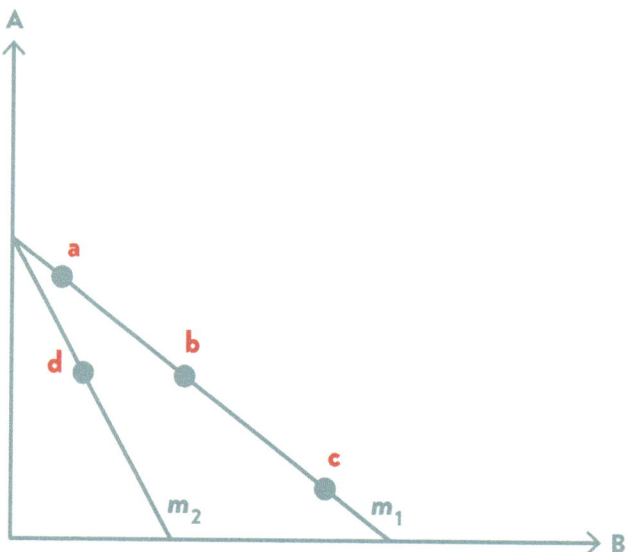

Fig. 1.2 Shift in the Budget Line. *Note* An increase in the price of good B causes the budget line to shift from m_1 to m_2. The budget line becomes steeper, with a lower maximum consumption of good B, while the maximum consumption of good A remains unchanged. Point d is one possible combination of the two goods along m_2.

the points a to d in the figure, she can now only afford d, but she can also choose any other point along the m_2 line.

Anna wonders: What happens if I earn more money, for example by working more? She draws Fig. 1.3 with a new budget line, m_3, which is a parallel shift of m_1. Now she thinks she would choose point e, but she can also afford any other combination of A and B along the new budget line.

We see that both price changes and income changes affect the choice set. But which point on the budget line will Anna choose? That depends on her preferences.

1.3 Preferences: Utility and Indifference Curves

Looking at Fig. 1.3, Anna feels she would be equally satisfied with points a, c, and d. It then logically follows that b must be better than these three. This is especially clear if we compare b and d: point b contains as much of good A as in point d, but more of good B—so it's better!

Anna's thoughts about the consumption combinations in Fig. 1.3 illustrate something very important about preferences in consumer theory. In fact, the theory assumes that most people think like Anna!

First, we assume consumers can always say, for any goods or combinations of goods, "I prefer this one to that one," or "I like the two equally." This is called *completeness*.

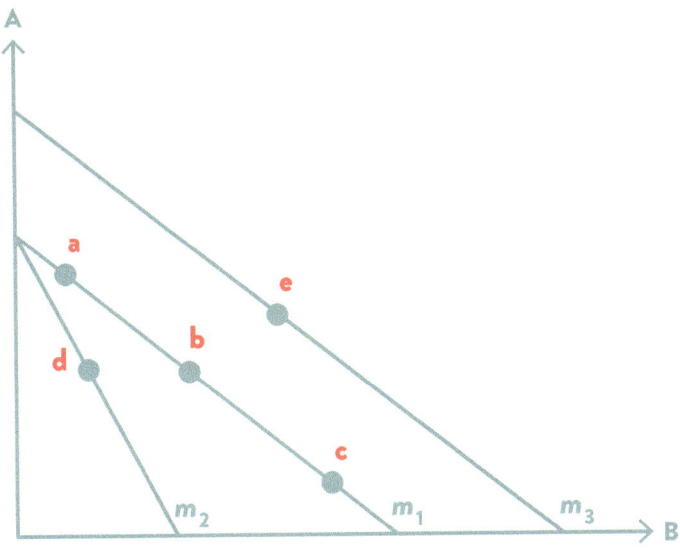

Fig. 1.3 Multiple Budget Lines. *Note* An increase in income causes a parallel shift of the budget line from m_1 to m_3. Point *e* is one possible combination of the two goods along m_3.

Second, we assume more is better than less. This is called *non-satiation*.

Third, we assume preferences are logically consistent: when Anna is equally satisfied with *a*, *d*, and *c*, and prefers *b* over *d*, she must also prefer *b* over *a* and *c*—anything else would be illogical! This is called *transitivity*.

So, when Anna is able to evaluate the different points *a, b, c,* and *d* (*completeness*); and she thinks *b* is better than *d* because *b* gives her more of one good and no less of the other (*non-satiation*); and this must mean that she also prefers *b* over *a* and *c* (*transitivity*), she has satisfied the three fundamental assumptions about preferences in consumer theory.

The satisfaction Anna derives from living well and enjoying herself is what economists call *utility*. And since economists find math both fun and useful, we often use a *utility function* to represent satisfaction. The utility function, $U(A, B)$, where U stands for utility, shows how much satisfaction the consumer derives from different combinations of goods A and B.

$$U = U(A, B) \quad \text{Utility function}$$

Assume that Anna's preferences can be represented by the following utility function:

$$U(A, B) = A^{0.5}B^{0.5} \quad \text{Balanced Cobb-Douglas utility function} \qquad (1.3)$$

This is an example of a Cobb-Douglas (CD) utility function, a functional form you might recognise from a previous course in mathematics or economics. It is named after the American economists Charles Cobb (1875–1949) and Paul Douglas (1892–1946). I call it a *balanced* CD-utility function because the exponents on the two goods in the function are identical (and sum up to one). We will often use the Cobb-Douglas form in our mathematical examples in consumer and producer theory, which make up the first two parts of this book. You will encounter a slightly more general version in Math Box 1.1, where the exponents are not necessarily equal.

As mentioned, Anna is equally satisfied with points *a*, *d*, and *c*. They give her the same utility. She draws a curve connecting all points that provide her with the same utility. We call this an *indifference curve*. The indifference curve passing through points *a*, *d*, and *c* is labelled U_2, as illustrated in Fig. 1.4.

Similarly, she draws the indifference curve U_1, which consists of combinations of the two goods that give her the same satisfaction as point *b*, and we know that the satisfaction here is higher than on U_2. Points along the indifference curve U_3 obviously yield even higher utility.

It is common to assume that indifference curves are convex to the origin. This is based on the assumption that most people prefer some balance in consumption. For example, Anna prefers *b* over *c* and *a*, and few would choose the extreme points on the budget line!

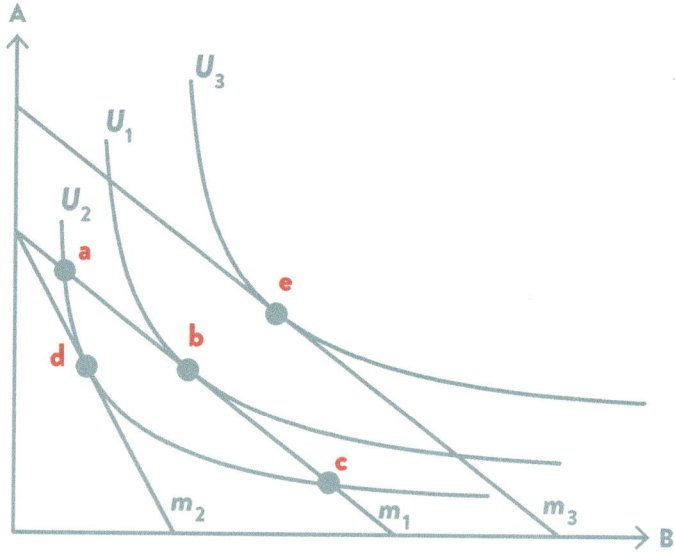

Fig. 1.4 Budget Lines and Indifference Curves. *Note* An indifference curve illustrates the consumer's preferences. All combinations of the two goods A and B along an indifference curve provide the consumer with the same utility. Combinations of A and B along the indifference curve U_1 give higher utility than those along U_2, while those along U_3 provide higher utility than both U_1 and U_2

In Math Box 1.1, we will become more familiar with the properties of indifference curves and take a closer look at the case of Cobb-Douglas preferences. At the end of the chapter, we examine other, somewhat more special cases.

The slope of the indifference curve plays a central role in consumer theory and has therefore been given a special name: the *marginal rate of substitution (MRS)*. It indicates how much of good A one is willing to give up to gain one more unit of good B, without changing overall utility. As we will soon see, understanding the MRS is essential to understanding consumer choice.

At this point, we need to introduce the concept of *marginal utility* (abbreviated *MU*). The marginal utility of good A tells us how much utility increases when consumption of A is increased by one unit:

$$\frac{\partial U}{\partial A} = MU_A \quad \text{Marginal utility of A} \tag{1.4}$$

Similarly, we can express the marginal utility of good B as:

$$\frac{\partial U}{\partial B} = MU_B \quad \text{Marginal utility of B} \tag{1.5}$$

To understand the relationship between marginal utility and the slope of the indifference curve, start with point a in Fig. 1.4. Here, the indifference curve is steep, which means that one is willing to give up quite a lot of A to gain one extra unit of good B. This means that the utility gain from one additional unit of B exceeds the utility loss from giving up one unit of A. Similarly, the indifference curve is quite flat at point c. This means the consumer is less willing to give up units of A in exchange for more B, because the marginal utility of B here is small relative to the marginal utility of A.

To describe the slope of the indifference curve mathematically, we can use the rule of implicit differentiation, which states that if you have a function $f(x, y) = c$, where c is a constant, then the following holds:

$$-\frac{dx}{dy} = \left(\frac{\partial f}{\partial y}\right) / \left(\frac{\partial f}{\partial x}\right) \quad \text{Implicit derivation rule}$$

With the utility function $U(A, B)$, where utility along an indifference curve is held constant—let's call it U_1—so that we start from $U(A, B) = U_1$, the rule of implicit differentiation tells us that:

$$-\frac{dA}{dB} = \left(\frac{\partial U}{\partial B}\right) / \left(\frac{\partial U}{\partial A}\right)$$

The expression at the numerator of the right-hand side is what we recognise as the marginal utility of B, MU_B, while the expression at the denominator is the marginal utility of A, MU_A. Thus, the slope of the indifference curve—that is, the marginal rate of substitution (MRS)—can be expressed as:

$$MRS = \frac{MU_B}{MU_A} \quad \text{Marginal rate of substitution} \tag{1.6}$$

A useful rule of thumb for remembering the MRS is that the numerator contains the marginal utility of the good on the first axis (the horizontal axis, in our case good B), while the denominator contains the marginal utility of the good on the second axis (the vertical axis, good A).

In Math Box 1.1 we discuss the slope of an indifference curve based on a specific utility function, which I think will give you a better understanding of how it works.

Math Box 1.1: The Indifference Curve

Assume the following utility function:

$$U(A, B) = A^\alpha B^{1-\alpha} \quad \text{General Cobb-Douglas utility function.}$$

I will refer to this as the *general* Cobb-Douglas utility function, since the exponents can take different values of α (the Greek letter alpha). The exponent α reflects how much weight the consumer places on good A in their consumption. If the consumer only cares about A, then $\alpha = 1$. If the consumer only cares about B, then $\alpha = 0$. If the consumer cares equally about both goods, then $\alpha = 0.5$, and we have a balanced Cobb-Douglas utility function.

To find the slope of the indifference curve for this utility function, we begin by deriving the marginal utility:

$$MU_A = \frac{\partial U}{\partial A} = \alpha A^{\alpha-1} B^{1-\alpha} \quad \text{Marginal utility of A}$$

Similarly for B:

$$MU_B = \frac{\partial U}{\partial B} = (1-\alpha) A^\alpha B^{-\alpha} \quad \text{Marginal utility of B}$$

The slope of the indifference curve is then given by:

$$MRS = \frac{MU_B}{MU_A} = \frac{(1-\alpha)A^\alpha B^{-\alpha}}{\alpha A^{\alpha-1} B^{1-\alpha}} = \frac{(1-\alpha)}{\alpha} \frac{A}{B}$$

We note that the larger the ratio A/B, the greater the MRS and the steeper the indifference curve (as in point a in Fig. 1.4), and the more of good A the consumer is willing to give up to obtain one more unit of good B. Conversely, the smaller the ratio A/B, the smaller the MRS and the flatter the indifference curve (as in point c in Fig. 1.4): in this situation, the consumer has only a small amount of A and is not particularly willing to give up more of it in exchange for more of good B. The Cobb-Douglas utility function, therefore, implies that the consumer wants to have a balanced consumption of the two goods—with the preferred balance depending on α—and this means that the indifference curve is convex toward the origin.

1.4 Utility Maximisation

In the discussion of Fig. 1.4, we saw that point b gives Anna higher utility than point a or c. But are there other combinations of the two goods along m_1 that she would prefer even more? The answer is no.

Figure 1.5 shows the budget line m_1 and four points a–d that lie within the feasible set. Point d cannot be the best choice because it lies inside the budget line: not all the money is being spent, and in our analysis, this is not a good idea (for now we disregard saving, but more on that in the next chapter).

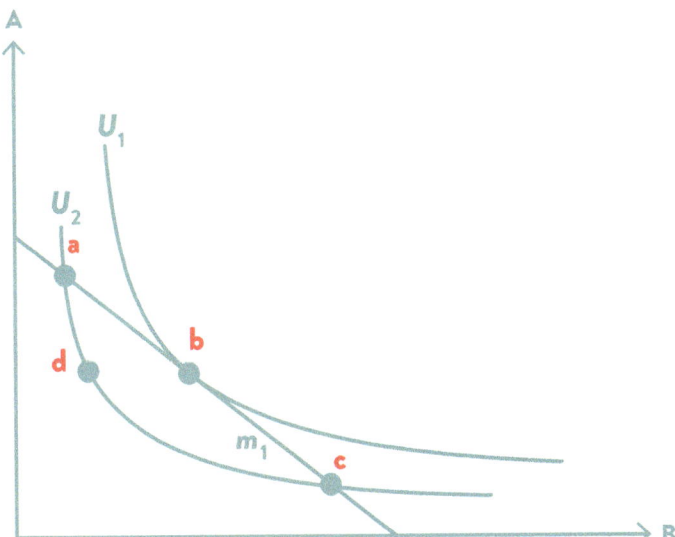

Fig. 1.5 Utility Maximisation. *Note* The consumer maximises their utility, given the feasible set, by choosing the combination of the two goods A and B where the indifference curve is tangent to the budget line—that is, where their slopes are equal. In the figure, this means the consumer will choose point b

At points a and c, the entire budget is spent, but the consumption is very unbalanced. Anna prefers a balanced consumption bundle, and we can see that point b gives her higher utility than the other two points on the budget line.

In fact, on the budget line m_1, point b gives Anna the highest utility. At this point, the indifference curve is tangent to the budget line, which means that the slopes of the two are equal:

$$MRS = \frac{MU_B}{MU_A} = \frac{p_B}{p_A} \quad \text{Optimal consumption rule} \qquad (1.7)$$

We now see why the MRS is so central in consumer theory: we need it to determine the consumer's choice. We can gain deeper insight into the condition for optimal consumption by reformulating the expression above:

$$\frac{MU_B}{p_B} = \frac{MU_A}{p_A} \quad \text{Equal bang for the buck} \qquad (1.8)$$

This means that at the optimum, the marginal utility per euro spent should be the same for both goods, or as they say in America: equal bang for the buck. If not, one could increase satisfaction by shifting money from the good that yields lower marginal utility per euro to the one that yields higher marginal utility per euro.

Here, we have used the insight from a graphical analysis—in our case, Fig. 1.5—to arrive at the mathematical formulation of the condition for optimal consumption, as shown in (1.7) and (1.8). From the figure, we established that the optimum is defined at the tangency point between the budget line and the indifference curve, and then we formulated the slopes of the two and set them equal.

In both consumer theory and producer theory, which we cover in the second part of the book, we will use this approach—what we might call the graphical approach—to find an optimum.

However, there is another way to find the optimum that you might recognise from a mathematics course, namely the Lagrange method. This method can be used for optimisation with a constraint, and utility maximisation is precisely such a problem: the consumer wants to maximise their utility $U(A, B)$ given the budget constraint $I = p_A A + p_B B$.

I will briefly go through this method here. The consumer's optimal choice between two goods A and B, for a given budget, can be found by first formulating the Lagrangian expression:

$$\mathcal{L} = U(A, B) - \lambda(p_A A + p_B B - I)$$

Here, λ (the Greek letter lambda) is the so-called Lagrange multiplier, which corresponds to a reformulated version of the budget constraint that must equal zero. We begin by maximising the Lagrangian expression with respect to A and B:

$$\frac{\partial \mathcal{L}}{\partial A} = \frac{\partial U}{\partial A} - \lambda p_A = 0$$

$$\frac{\partial \mathcal{L}}{\partial B} = \frac{\partial U}{\partial B} - \lambda p_B = 0$$

Note that the first term on the right-hand side of the two first-order conditions is the marginal utility, namely MU_A and MU_B, respectively. By combining these two expressions, we find the condition for optimal consumption as described in (1.7), which is:

$$MRS = \frac{MU_B}{MU_A} = \frac{p_B}{p_A}$$

We then differentiate the Lagrangian expression with respect to λ and set it equal to zero:

$$\frac{\partial \mathcal{L}}{\partial \lambda} = p_A A + p_B B - I = 0$$

This gives us the budget constraint, namely that $I = p_A A + p_B B$, and based on the three Lagrange first-order conditions, we can thus derive the same condition for optimal consumption as with the graphical approach. The two methods are therefore equivalent. Although it is not necessary to use the Lagrange method to find the optimum, it is useful to be familiar with it.

To better understand what lies behind the condition for optimal consumption described in Eq. (1.7), alternatively formulated as in (1.8), it is useful to work with a concrete utility function and explicitly show how the utility-maximising consumption depends on prices and income. We do this in Math Box 1.2.

The path to the optimum involves several steps—some mathematical ingredients—and it can be useful to have a recipe before starting the work. Here it is:

Three steps to consumer optimum:

▶ **Recipe Number 1: Three Steps to Consumer Optimum**

Step 1: Compute the MRS. Use the utility function to find the MRS, that is, the ratio between the marginal utilities of the two goods.
Step 2: Apply the Tangency Condition. Then insert this expression for the MRS into the condition for optimal consumption: $MRS = pB/pA$. From this, we can find the optimal consumption ratio, A/B.
Step 3: Solve for each good. Solve the optimal consumption ratio for one of the goods, and substitute this expression into the budget constraint. From this, we can find the optimal consumption of one good, which can then be inserted back into the budget constraint to find the optimal consumption of the other good.

Math Box 1.2: Utility Maximisation

Assume the general Cobb-Douglas utility function:

$$U(A, B) = A^\alpha B^{1-\alpha}$$

Step 1 asks us to start with the specific utility function and find the MRS. From Math Box 1.1, we know that the marginal rate of substitution is given by:

$$MRS = \frac{(1 - \alpha)}{\alpha} \frac{A}{B}$$

Step 2 states that we should use this expression for the MRS in the optimality condition described in (1.7):

$$MRS = \frac{(1 - \alpha)}{\alpha} \frac{A}{B} = \frac{p_B}{p_A}$$

This can be expressed as:

$$\frac{A}{B} = \frac{\alpha}{(1 - \alpha)} \frac{p_B}{p_A}$$

Which is the optimal consumption combination of the two goods. We have now completed step 2.

Step 3 first asks us to solve the optimal consumption combination for one of the goods. We solve for good A and find:

$$A = \frac{B\alpha}{(1 - \alpha)} \frac{p_B}{p_A}$$

We insert this into the budget constraint $I = p_A A + p_B B$, and get:

$$I = p_A \left(\frac{B\alpha}{(1 - \alpha)} \frac{p_B}{p_A} \right) + p_B B$$

A bit of manipulation gives us:

$$(1 - \alpha)I = \alpha p_B B + (1 - \alpha)p_B B$$

This results in the following expression for B:

$$B = \frac{(1 - \alpha)I}{p_B} \quad \text{Optimal consumption of B}$$

We can then substitute the optimal consumption of B into the budget constraint:

$$I = p_A A + p_B \left(\frac{(1 - \alpha)I}{p_B} \right)$$

And we find that:

$$A = \frac{\alpha I}{p_A} \quad \text{Optimal consumption of A}$$

Voilà! This is the utility-maximising demand for the two goods given the specified utility function. We see that the demand for A increases with α and income I, and decreases with the price p_A. This makes sense: a higher weight on A in the utility function and a higher income makes the consumer buy more of A, while a higher price of A reduces the demand for this good. Similarly for B: a higher weight on B in the utility function and higher income increase the demand for B, whereas a price increase for B works in the opposite direction.

1.5 Other Types of Preferences

So far, we have studied Anna's choice between her home base and other goods. From Math Box 1.2, we see that since Anna assigns equal weight to the base and other goods in her preferences ($\alpha = 0.5$), she will spend an equal amount of money—half of her income—on each of the two goods: $p_A A = p_B B = 0.5I$. For example, if the rent doubles, she would want a flat half the size, so that $p_B B$ remains unchanged.

This applied to the choice between her home base and other goods. But for other choices, different preferences may apply. And different people may of course have different preferences.

Anna is attending a lecture in microeconomics, and the professor is talking about peppers—green and yellow peppers. Many would think it does not matter whether the pepper is green or yellow, at least the professor thinks so, and in such cases, they would be perfect substitutes. The utility function in this case can be formulated as:

$$U(A, B) = A + B \quad \text{Perfect substitutes}$$

Here, A and B stand for yellow and green peppers. The indifference curve in this case will be linear, as shown by U_{Sub} in Fig. 1.6, panel A. With perfect substitutes, the marginal rate of substitution is constant. The professor says he is willing to trade one green pepper for one yellow pepper, even if it is his last green one!

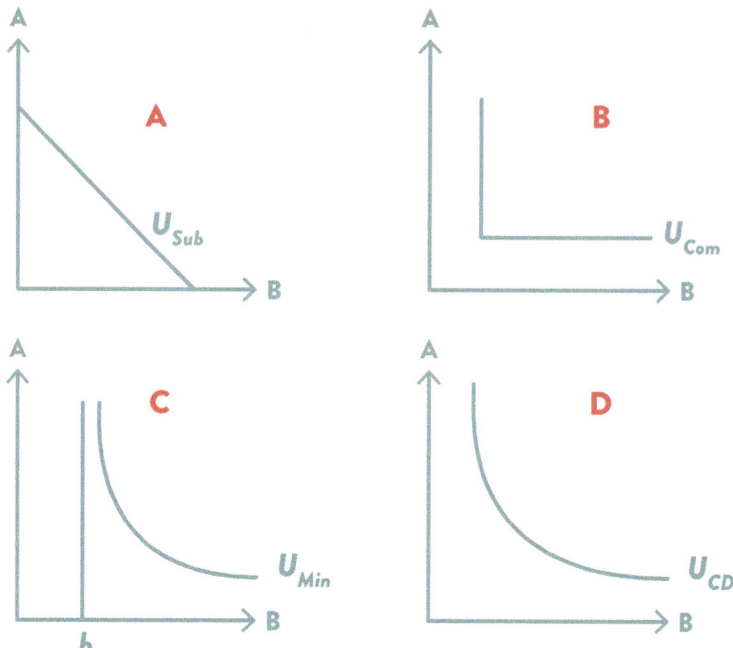

Fig. 1.6 Indifference Curves with Different Preferences. *Note* When the goods are perfect substitutes, the indifference curve is linear, as shown by U_{Sub} in panel A. When the goods are perfect complements, the indifference curve is L-shaped, as shown by U_{Com} in panel B. If B is a basic good with minimum consumption b, we get an indifference curve as shown in panel C. And in panel D, we have a standard Cobb-Douglas indifference curve

This contrasts with Cobb-Douglas preferences, where the willingness to substitute one good for another depends on the starting point: You're more willing to give up something you have plenty of than something you have only a little of.

The other extreme is perfect complements, where there is zero possibility of substitution. Imagine you have one right shoe and one left shoe: Would anyone want to trade their right shoe for an extra left shoe? Such goods are called perfect complements, and the utility function is then given by:

$$U(A, B) = min(A, B) \quad \text{Perfect complementarity}$$

The indifference curve is then given by U_{Com} in Fig. 1.6, panel B.

Some goods can be classified as basic goods—we need a minimum consumption of these before we can start thinking about spending money on other things. This could mean having a roof over your head or just enough calories to get by. Here is an example of a utility function where B is a basic good with a minimum consumption b:

$$U(A, B) = A^{0.5}(B - b)^{0.5} \quad \text{B as basic good}$$

As you can see, this is a variant of a balanced Cobb-Douglas function, but with the twist, that consumption of B must be above the minimum level b before one can derive any utility from any good. U_{Min} in Fig. 1.6, panel C, shows an indifference curve with such a basic good.

And finally, in panel D, I have included a standard Cobb-Douglas indifference curve for comparison.

1.6   How Does a Price Change Affect Consumption?

Anna is at work thinking about today's microeconomics lecture. The lecturer talked about how a price change affects consumers' choices. He discussed both how a price change for a good affects the demand for that good, known as the own-price effect, and how it affects the demand for other goods, known as the cross-price effect.

The professor teaches microeconomics

The professor explained that the own-price effect is usually straightforward: a higher price for a good leads to a decrease in demand for that good, and a lower price leads to higher demand. However, he also said that this is not necessarily always the case and promised to return to this topic in the next lecture (see the next chapter).

On the other hand, the cross-price effect is less certain. Anything could happen, depending on consumers' preferences. He found this very interesting and waved his pointer stick enthusiastically.

The professor ended the lecture with the following question: What is the cross-price effect under Cobb-Douglas preferences? He left the question hanging and asked the students to think about it until the next lecture.

Anna was inspired by the lecture, and during her afternoon shift at the grocery shop, she observed what customers had in their shopping carts. This week, bacon was on sale, and she noticed that alongside a lot of bacon, more eggs than usual were also being sold. However, she barely saw a single customer buying pancetta (an unsmoked Italian bacon).

What Anna observed was a positive own-price effect for bacon (bacon on sale, more demand for bacon), a positive cross-price effect for eggs, but a negative cross-price effect for pancetta. Eggs and bacon tend to go together (they are complements), while bacon and pancetta are more similar (in most recipes you can use either one or the other—they are substitutes).

Anna thought about the different types of preferences the professor drew on the board, as shown in Fig. 1.6. With perfect complementarity, the indifference curve is L-shaped. Eggs and bacon don't fit together quite as perfectly as a right and left shoe, but the indifference curve is probably more L-shaped than the Cobb-Douglas function. And bacon and pancetta may not be perfect substitutes, but their indifference curve is likely to be nearly linear.

In a quiet moment behind the cash register, Anna drew two figures on a napkin. Figure 1.7 represents the choice between bacon and eggs, while Fig. 1.8 shows the choice between bacon and pancetta.

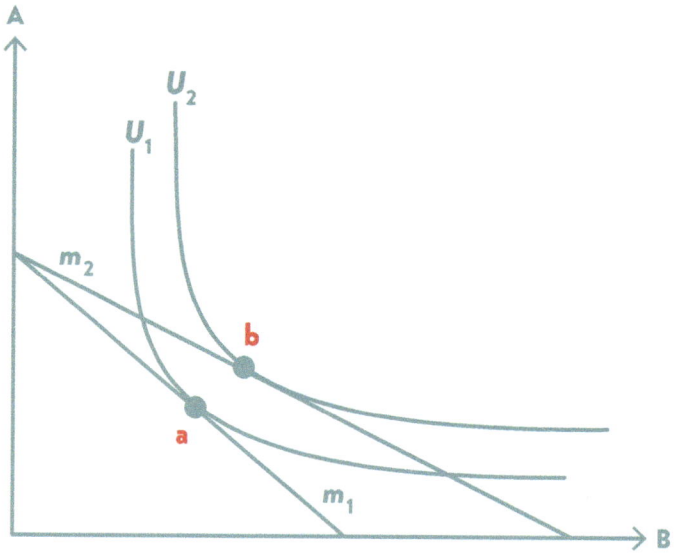

Fig. 1.7 Complements. *Note* A lower price for good B has caused the budget line to rotate from m_1 to m_2. The consumer's choice has here changed from point *a* to point *b*, and we notice that the consumption of A has increased. When the demand for a good increases as the price of the other good falls (and vice versa), we call the goods complements. We can think of good A as eggs and B as bacon: a lower price for bacon has led to higher demand for eggs. Eggs are thus complementary to bacon

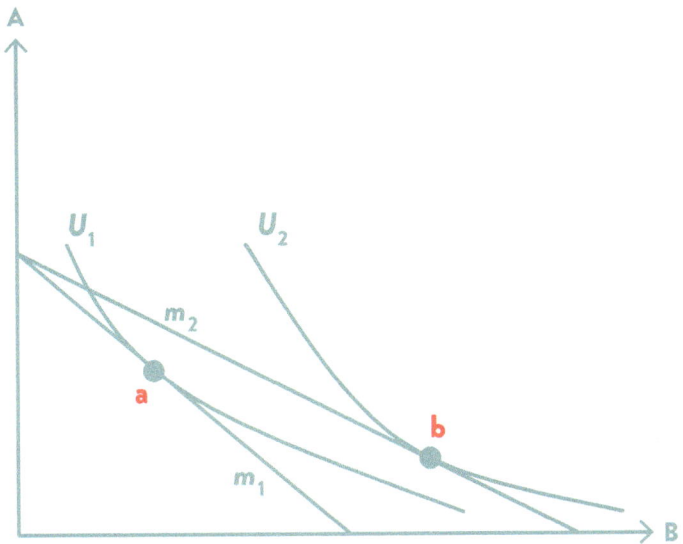

Fig. 1.8 Substitutes. *Note* A lower price for good B has caused the budget line to rotate from m_1 to m_2. The consumer choice has here changed from point a to point b, and we notice that the consumption of A has decreased. When the demand for a good decreases as the price of the other good falls (and vice versa), we call the goods substitutes. We can think of good A as pancetta and B as bacon: a lower price for bacon has led to lower demand for pancetta. Pancetta is thus a substitute for bacon

The starting point in both figures is point a, and with a lower price on bacon (B), the budget line shifts to m_2. This moves us to point b. With the more L-shaped indifference curve in Fig. 1.7, we see that demand for A (eggs) increases, while with the more linear indifference curve in Fig. 1.8, demand for A (pancetta) decreases.

Anna is satisfied with her analysis. The theory she has learned has made it easier for her to understand consumer choices. She looks forward to the next lecture, where they have been promised more on consumer choices and some real theoretical treats!

She also thinks about the question the professor gave them at the end of the class about the cross-price effect with Cobb-Douglas preferences. Hmmm… okay, here comes yet another customer with eggs and bacon—she'd better think more about cross-price effects back at her flat!

1.7 Summary

In this chapter, we have studied the consumer's choice between two goods. The options are limited by the money available and the prices of the different goods. We assume that consumers do their best given these constraints, which means they maximise their utility. We have examined how changes in the price ratio between goods and changes in income affect the choice set and the consumer's choice.

We have also studied cross-price effects—that is, how a price change in one good affects the demand for the other—and shown that this depends on whether the consumer perceives the goods as substitutes or complements.

By the way, have you found the answer to the question about the cross-price effect under Cobb-Douglas preferences? If not, just keep reading. The solution is in the next chapter!

1.8   Key Terms

Budget line: The budget line shows all possible combinations of the two goods that the consumer can afford.

Utility function: Expresses the satisfaction or happiness one derives from consuming different combinations of goods.

Indifference curve: Shows combinations of two goods that provide the same level of utility.

Marginal utility: The increase in utility that results from consuming one additional unit of a good.

The marginal rate of substitution (MRS): Shows the slope of the indifference curve, that is, how much the consumer is willing to reduce good A in exchange for one extra unit of good B without changing their overall utility.

Perfect substitutes: When one good can completely replace the other. The indifference curve is linear.

Perfect complements: Goods that naturally go together. The indifference curve is L-shaped.

Own-price effect: How the demand for a good is affected by a change in the price of that same good.

Cross-price effect: How the demand for a good is affected by a change in the price of another good.

1.9   Multiple-Choice Exercises

1.1: Budget Line

Assume that the price of B is unchanged. What must have happened when we move from m_1 to m_2 in the figure below?

A. Higher income, higher price of A
B. Lower income, lower price of A
C. Unchanged income, lower price of A
D. Higher income, lower price of A

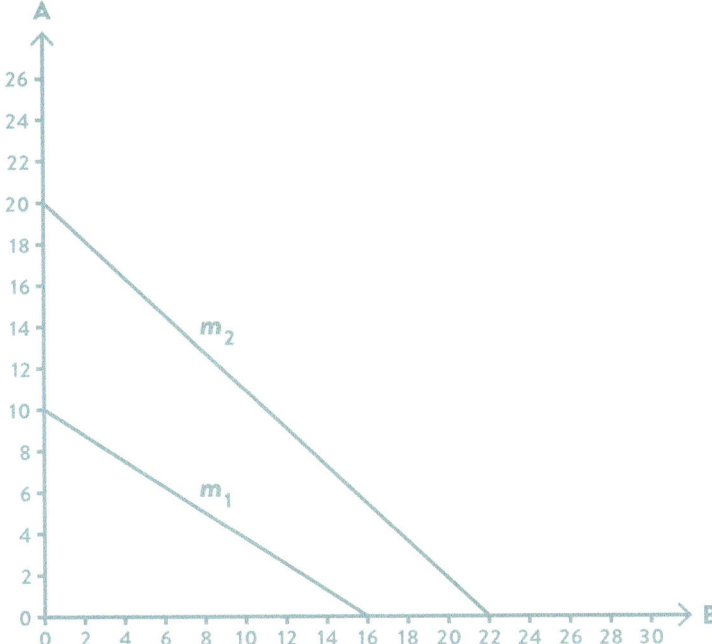

1.2: The Marginal Rate of Substitution

Consider the figure below. What is true?

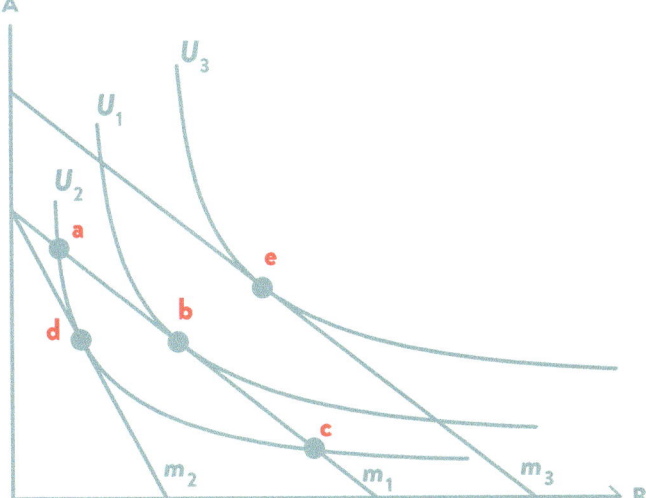

1. MRS is the same at points *b* and *e*.
2. MRS is higher at point *a* than at point *b*.

A. Neither 1 nor 2
B. Both 1 and 2
C. Only 1
D. Only 2

1.3: Utility Maximisation

Assume that the price ratio between goods A and B is given by $(p_B/p_A) = 2.5$ and that the consumer has a consumption bundle where $MRS = (MU_B/MU_A) = 3$. To maximise utility, the consumer should:

A. Buy less B and more A
B. Buy more B and less A
C. Not change their consumption
D. Buy more B and more A

1.4: Substitutes or Complements?

Consider the change in the consumer's choice from point *a* to point *b* in the figure below.

Which of the following statements is true?

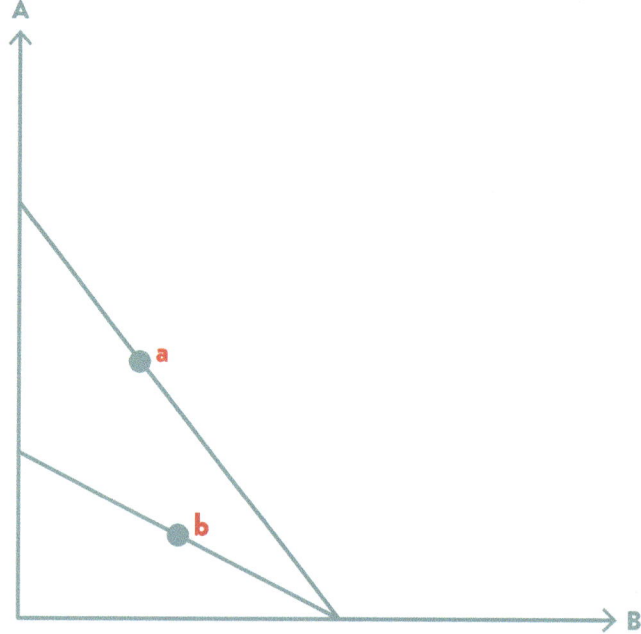

A. The consumer has received a higher income. A and B are substitutes
B. The consumer has received a higher income. A and B are complements
C. The consumer has not received a higher income. A and B are substitutes
D. The consumer has not received a higher income. A and B are complements

Solutions: 1.1 D; 1.2 B; 1.3 B; 1.4 C

More About Consumer Choice

2

Why Brian buys more frozen pizza when the price goes up: "It was the only thing I could afford,"
he says.

And why an interest rate increase can be good news for some and bad news for others.

2.1   Introduction

We ended the previous chapter by looking at the cross-price effect, and we saw
that it could be either positive or negative, depending on preferences.

Brian likes beer, preferably an IPA from the local microbrewery

© The Author(s), under exclusive license to Springer Nature Switzerland AG 2026   25
K. Bjorvatn, *Microeconomics Made Simple*, Classroom Companion: Economics,
https://doi.org/10.1007/978-3-032-06354-0_2

Have you figured out what the cross-price effect is under Cobb-Douglas preferences?

Answer: It's zero! With these preferences, a change in the rent does not affect the demand for other goods. And a change in the price of other goods does not affect the choice of your home base.

To understand why, we will take a closer look at the effect of a price change. We will see that it can be split into two sub-effects: the substitution effect and the income effect.

We will also look more closely at how a change in income affects consumption. One might assume that higher income necessarily leads to higher consumption, but that is not always the case. We will divide goods into two types: normal goods and inferior goods.

In this chapter, we also get to know another of the book's main characters: Brian.

He is also in his first year at the School of Economics, is interested in computers, good at maths, and fond of beer and frozen pizza. Brian's story helps us shed light on key topics in consumer theory—namely, inferior goods, Giffen goods, and fungibility.

The chapter ends with an analysis of saving, which can be seen as a trade-off between consumption today and in the future. Anna realises that she will be very busy next semester and won't be able to work as much at the grocery shop. To smooth her consumption over time, she wants to save. How does a change in the interest rate affect her consumption possibilities?

And what if she had taken out a loan instead? We will see that an increase in interest rates is good news for savers and bad news for borrowers.

2.2   Substitution Effect and Income Effect

When the price of a good changes, two things happen to the budget line: its slope changes, and the choice set changes. We saw this in Chapter 1—for example, in Fig. 1.2, where an increase in the rent rotates the budget line from m_1 to m_2. The budget line becomes steeper, and the choice set—the area under the budget line—shrinks.

The change in the slope of the budget line reflects a change in relative prices: in Fig. 1.2, the home base has become relatively more expensive. The change in consumption resulting from this is called the substitution effect.

The shrinking of the choice set is due to the loss of purchasing power: your income doesn't stretch as far when housing has become more expensive. The change in consumption resulting from this reduced purchasing power is called the income effect.

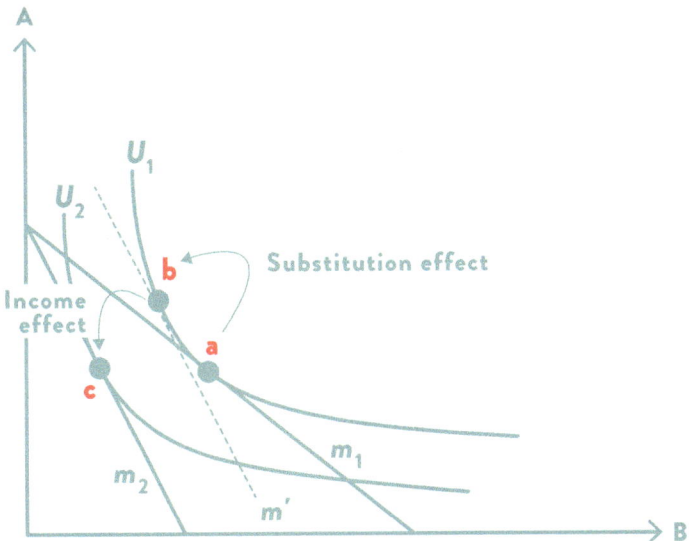

Fig. 2.1 Substitution effect and income effect. *Note* The figure shows how a price increase of good B, which rotates the budget line from m_1 to m_2, affects the demand for the two goods A and B. Initially, the consumer's choice is at point *a*, while the choice after the price increase is at point *c*. We use the helper line m' which is a parallel shift of the new budget line up to the original indifference curve, to find the substitution effect, indicated by the movement from *a* to *b*. The income effect is the movement from *b* to *c*

We illustrate the two effects of a price change in Fig. 2.1. Here we look at the effect of an increase in the rent, which rotates the budget line from m_1 to m_2. To isolate the substitution effect, we ask: what happens to demand when we consider only the change in relative prices?

The thought experiment here is to ignore the impact of the price change on purchasing power by compensating the consumer with enough income to bring her back to her original utility level. She now has the same utility as before but achieves it by consuming more of the good that has become relatively cheaper and less of the one that has become relatively more expensive.

We illustrate this with the dashed helper line m' which is a parallel shift of the new budget line m_2 up to the original indifference curve U_1, with the tangency point at *b*. The substitution effect is thus the movement from point *a* to point *b*.

The income effect captures the impact on demand caused by the change in the choice set when the price of a good changes. In the figure, this is shown by the parallel shift from the dashed helper line to the new budget line. The income effect is thus the movement from point *b* to point *c*.

Now we also have the answer to why the cross-price effect is zero with Cobb-Douglas preferences: a higher rent causes a positive substitution effect for other goods (since other goods become relatively cheaper when the rent increases), but this positive effect is exactly cancelled out by the negative income effect (higher rent means lower purchasing power). In total, the demand for other goods is not affected by a change in the rent—the cross-price effect is zero.

In Math Box 2.1, we show this result more formally. This also involves some mathematical steps, and as in the previous chapter, it can be useful to have a recipe to reach the conclusion. The new recipe begins where the previous one ended.

▶ **Recipe Number 2: Three Steps to the Substitution Effect and Income Effect**

Step 1: Find the initial utility level. Insert the optimal consumption of A and B into the utility function and find the utility level before the price change.

Step 2: Derive the substitution effect. Use the optimal consumption combination to find the ratio between the consumption of the two goods at the new price ratio. Solve for one of the goods, insert this into the utility function, and keep the utility level the same as before the price change. From this, we can find the optimal consumption of one good, and then by using the utility function, find the optimal consumption of the other good as well: we have now found the substitution effect.

Step 3: Derive the income effect. Use the optimal consumption combination, solve for one of the goods, insert into the budget constraint, and find the optimal consumption with the new prices. The difference between the consumption found in step 2 and the consumption found here is the income effect.

Math Box 2.1: Substitution Effect and Income Effect

Step 1 on the way to finding the substitution effect and income effect is about determining the utility level before the price change. Assume the balanced Cobb-Douglas utility function:

$$U(A, B) = A^{0.5}B^{0.5}$$

Math Box 1.2 shows the optimal consumption combination and the optimal consumption of the two goods for the general Cobb-Douglas utility function $U(A, B) = A^{\alpha}B^{1-\alpha}$. By setting $\alpha = 0.5$, we find the corresponding expressions for the balanced Cobb-Douglas utility function as:

$$A = \frac{0.5I}{p_A} \quad \text{Optimal consumption of A, balanced CD}$$

$$B = \frac{0.5I}{p_B} \quad \text{Optimal consumption of B, balanced CD}$$

Now assume that $I = 1$ and that the initial prices are $p_A = p_B = 1$. We then see that the optimal consumption of the two goods initially, which we denote with subscript a, is $A_a = B_a = 0.5$, which inserted into the utility function results in:

$$U(A_a, B_a) = U_a = 0.5^{0.5}0.5^{0.5} = 0.5 \quad \text{Utility before price change}$$

With this, we have found the utility level before the price change, $U_a = 0.5$, and have now completed step 1.

Step 2 is about finding the substitution effect, the movement along the utility level U_1 from point a to point b in Fig. 2.1. This is the most complicated step, but the key is that we move along the original utility level found in Step 1, so this must remain fixed.

Assume that the price of good B increases to $p_B = 2$. With prices $p_A = 1$ and $p_B = 2$, we find from the optimal consumption combination that $A = 2B$. That is, with the new price ratio, the consumer will always choose to consume twice as much of A as B. We insert this expression for A into the utility function while keeping utility constant at $U = 0.5$:

$$U = A^{0.5}B^{0.5} = 0.5 \Rightarrow (2B)^{0.5}B^{0.5} = 0.5$$

This can be solved for B, giving us the demand for B at the tangency point between the dashed helper line and the indifference curve, which we denote with subscript b, as follows:

$$B_b = \frac{0.5}{2^{0.5}} \approx 0.35$$

We can insert this into the utility function while keeping the utility level fixed, and find that:

$$A^{0.5}(0.35)^{0.5} = 0.5$$

From this expression, we can find the demand for A at the same tangency point as:

$$A_b = \frac{0.5^2}{0.35} \approx 0.71$$

The substitution effect (SE) can thus be summarised as:

$$A_a(0.5) \rightarrow A_b(0.71)SE_A \approx 0.21$$

$$B_a(0.5) \rightarrow B_b(0.35)SE_B \approx -0.15$$

The substitution effect causes an increase in A from 0.5 to 0.71, and a decrease in B from 0.5 to 0.35. We have now completed step 2.

Step 3 asks us to find the choice on the new budget line, which we denote with subscript c. This is straightforward. Using the formula for optimal consumption shown above, we can find the optimal consumption after the price change, that is, when the price of good B increases to $p_B = 2$, as $A_c = 0.5, B_c = 0.25$. This is the consumption combination at point c in Fig. 2.1.

The income effect (IE) can therefore be summarised as:

$$A_b(0.71) \rightarrow A_c(0.5)IE_A \approx -0.21$$

$$B_b(0.35) \rightarrow B_c(0.25)IE_B \approx -0.1$$

The income effect leads to a reduction in A from 0.71 to 0.5, and a reduction in B from 0.35 to 0.25.

We see in Fig. 2.1 that the substitution effect pulls consumption toward more of the other good when the rent increases. This will always be the case: the substitution effect means increased consumption of the good that has become relatively cheaper. The income effect, on the other hand, is less predictable, as we will soon see. In Fig. 2.1, the income effect pulls consumption toward lower levels of both goods, but this is not always the case.

2.3   Normal and Inferior Goods

In the previous chapter, we saw that when Anna's income increases, she wants a bigger flat and more of other goods. Economists call these normal goods. This is illustrated in Fig. 2.2. The increased income causes a parallel shift in the budget line from m_1 to m_2, and Anna chooses to move from point a to point b, an increase in consumption of both goods. The same would be true for any choice on m_2, along the line segment between points c and d: increased income leads to higher demand for both goods, meaning they are normal goods.

We can draw a line connecting the consumption points as income increases; this line is called the income expansion line.

But now it's time to get to know Brian better. He is also a first-year student at the School of Economics, but he has a somewhat different lifestyle than Anna. He

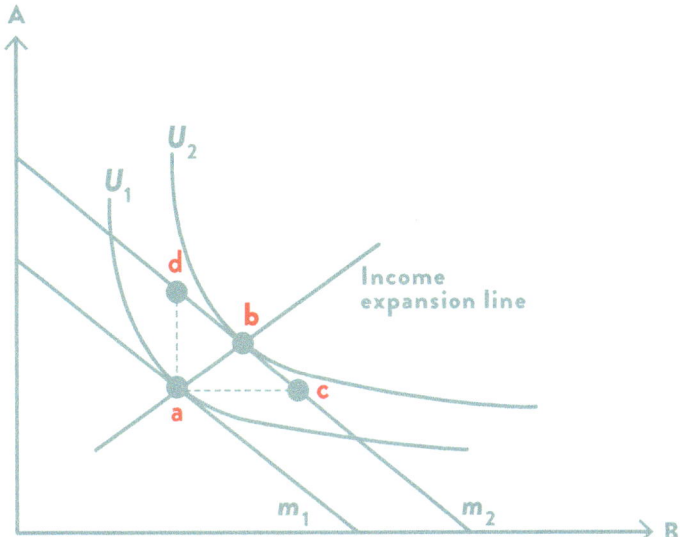

Fig. 2.2 Normal Goods. *Note* An increase in income causes a parallel shift in the budget line from m_1 to m_1. In the figure, this results in a change in consumption from point a to point b, and we see that the consumption of both goods increases as a result of the higher income. The same applies for any choice on m_2 in the interval between points c and d. These are called normal goods

finds it more comfortable to ask his mother for money than to earn it himself, and mostly lives on frozen pizza, especially when he's short on cash, as we shall now see.

One day, while Anna is at the checkout, Brian comes along with a shopping basket represented by point b in Fig. 2.3. In this case, good B is the pizza, and A represents other things (including a few bottles of the local IPA, which he insists are essential for washing it down). This surprises her a bit, since the week before Brian had chosen a shopping basket represented by point a, while complaining about being short of money. Now he's received money from his mother, which gives him the budget line m_2 and he chooses point b. What surprises Anna?

What surprises Anna is that Brian buys *less* pizza now that he has more money. Goods that we buy less of when we get better off are called inferior goods. Not because they are not valued by the consumers—quite the opposite: these are the goods we turn to when we are short on cash. But when income rises, we choose other things instead. In Fig. 2.3, any consumption choice on budget line m_2 outside the interval between points c and d means one of the goods is inferior: to the left of d, B is inferior; to the right of c, A is inferior.

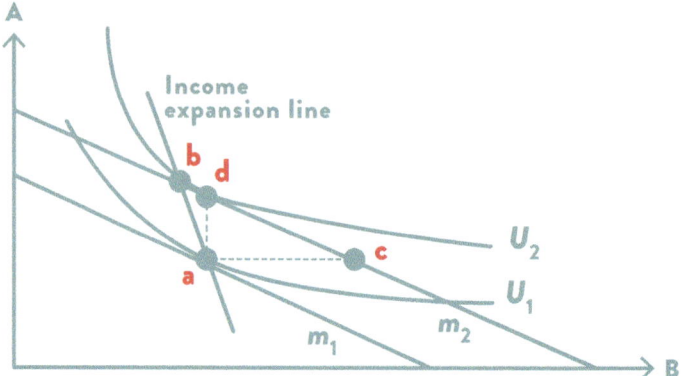

Fig. 2.3 Inferior Goods. *Note* An increase in income causes a parallel shift in the budget line from m_1 to m_2. In the figure, this leads to a change in consumption from point a to point b, where we see that while the consumption of good A has increased, the consumption of good B has decreased. Thus, B is an inferior good. Any choice on m_2 outside the interval between points c and d means that one of the goods is inferior: to the left of d, B is inferior; to the right of c, A is inferior

2.4   Giffen Goods

Brian's shopping basket continues to be a source of surprises. Once, Anna noticed he bought more frozen pizza during a week when the price was unusually high (it was often on sale, but not this week). Had Brian misread the price? No, Brian explained that because the pizza was so expensive, he couldn't afford much else!

Brian shopping, Anna at the counter

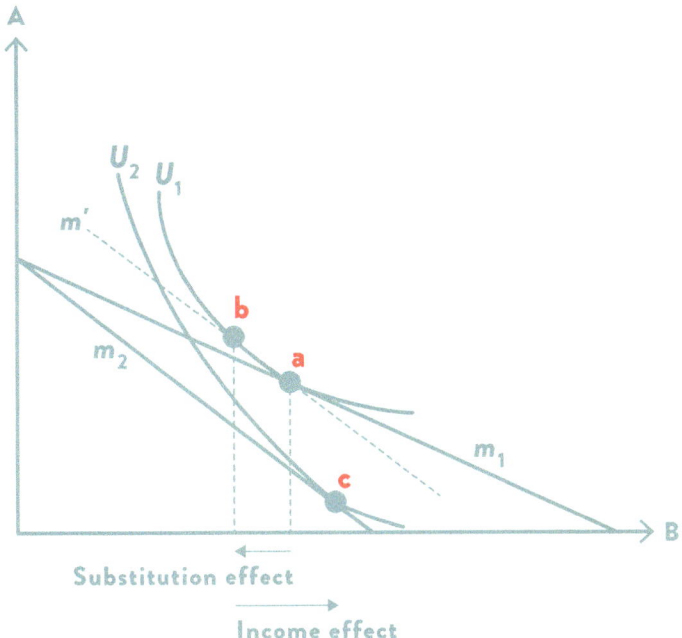

Fig. 2.4 Giffen Good. *Note* The price increase of good B causes the budget line to rotate from m_1 to m_2. The substitution effect is the movement from point a to point b, while the income effect is the movement from point b to point c. We see that B is an inferior good because the lower real income (caused by the higher price of B) increases the demand for this good. Since in this case the income effect dominates the substitution effect, we call B a Giffen good, where a higher price leads to higher demand

We analyse this in Fig. 2.4, where the high price of the basic good causes the budget line to shift from m_1 to m_2, which is steeper than m_1. As discussed earlier, the substitution effect of the price increase is the movement from point a to b: a higher price for the basic good leads to lower consumption of it. This is as expected—we can rely on the substitution effect!

The income effect, however, is less predictable. In the figure, the income effect is the movement from b to c, showing higher demand for the basic good. We've learned that this must be because it's an inferior good: lower income leads to higher consumption. What is special in this case is that the income effect dominates the substitution effect. Frozen pizza for Brian is not just an inferior good—it is a *strongly* inferior good!

A good whose demand goes up if the price goes up is called a Giffen good, named after the Irish priest Robert Giffen, who wrote about the potato famine in Ireland during the 1840s. The term "Giffen goods" can be attributed to the British economist Alfred Marshall (1842–1924), who in his textbook *Principles of Economics* from 1895 wrote:

> As Mr. Giffen has pointed out, a rise in the price of bread makes so large a drain on the
> resources of the poorer labouring families and raises so much the marginal utility of money
> to them, that they are forced to curtail their consumption of meat and the more expensive
> farinaceous foods: and, bread being still the cheapest food which they can get and will take,
> they consume more, and not less of it.

A higher price on bread led poor working-class families to cut back on meat and
other expensive items, while increasing their consumption of bread, since it was
still relatively cheap. Economists have long searched for other and more current
examples of Giffen goods, and recently American researchers managed to docu-
ment this in China, where they found that a price increase in rice led to increased
demand for rice.[1] And now, in this book, we have documented that the same
applies to Brian and frozen pizza! Although Giffen goods are not a common phenomenon, they are certainly an interesting theoretical possibility—a theoretical
treat, one might say.

2.5   Fungibility

Brian's mother, Audrey, is worried about her son. She thinks he eats too little
and too poorly, and wants to help him, but she is hesitant to give him money
because she has seen that he mostly spends it on beer. Therefore, she decides to
buy food for him instead. Figure 2.5 shows what happens.

Here, we can think of B as basic commodities such as food and A as other
things, such as (Indian Pale) Ale. Brian's budget line before the gift is given is
m_1 and his initial choice is at point a. With the grocery bag from his mother,
represented by B_0 he could reach point c, which is what his mother wants. But
what does Brian do?

The gift from his mother expands his choice set. The new budget line now follows m_2, but with a twist. If Brian spends all his own money on Ale, the maximum
point remains the same as before, where m_1 intersects the vertical axis (we assume
Brian cannot trade away the food from his mother for additional beer).

We see that Brian chooses point b, increasing consumption of both goods. He
thanks his mother for the food but still spends more money on his favourite drink!
This is called fungibility: earmarked gifts lead to rearrangements of one's own
funds, illustrated in the figure by a movement from point a to d along m_1. This
makes the effect of the gift smaller than one might initially expect.

---

[1] Robert T. Jensen & Nolan H. Miller (2008). Giffen behavior and subsistence consumption, American Economic Review 98 4: 1553–1577.

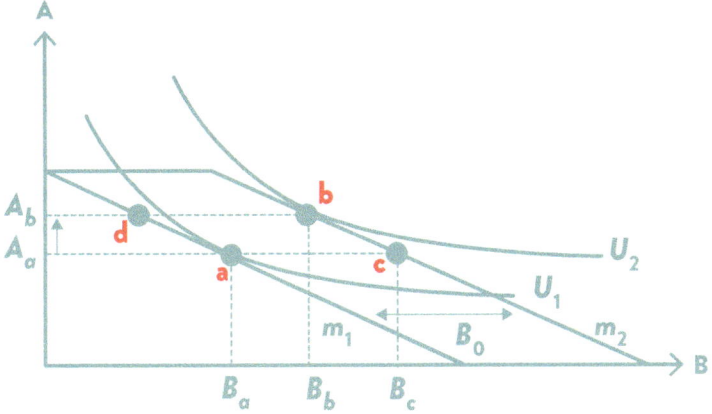

Fig. 2.5 Fungibility. *Note* A gift in the form of good B, B_0, shifts the budget line from m_1 to m_2. Note that the new budget line has a kink and therefore crosses the vertical axis at the same point as before: the gift is earmarked for B and does not increase the maximum consumption of good A. If the consumer does not reallocate any of their own money, we move from a to c. But from the figure we see that the optimal choice is at point b, with higher consumption of both goods. The consumer has achieved this by reallocating their own resources, which can be seen from the movement from a to d on the original budget line, m_1. This is called fungibility

Audrey decides to buy food for Brian

2.6   Saving

Anna is approaching the end of her studies at the School of Economics and sees that next semester will require more schoolwork, and therefore less time for the job at the grocery shop. Therefore, she sets up a saving plan.

We divide time into two periods, which for Anna means the last two semesters of her studies. Her job at the grocery shop gives her an income I_1 in period 1 and I_2 in period 2. She uses this income to consume x_1 in the current period and x_2 in the next.

Here, we have combined everything Anna enjoys related to her home base and leisure activities and simply call it consumption. We assume that the price of this consumption basket is the same in both periods, and for simplicity, we say the price is equal to 1.

The budget constraint in period 1 can be written as:

$$x_1 = I_1 - s \quad \text{Budget constraint period 1} \tag{2.1}$$

If $s > 0$, she spends less than she earns in period 1, meaning she saves. But she can also choose to spend more than she earns in this period, and in that case $s < 0$, meaning she takes out a loan. Saving yields a return, and correspondingly, one must pay interest on a loan. For simplicity, we assume that the interest rate on deposits and loans is the same and call it r. The budget constraint in period 2 is:

$$x_2 = I_2 + s(1 + r) \quad \text{Budget constraint period 2} \tag{2.2}$$

If Anna saves money in period 1, her consumption in period 2 can exceed her income in that period. But if she borrows money, she must repay the loan (with interest), and then her consumption in period 2 will necessarily be lower than her income in that period. We can divide Eq. (2.2) by $(1 + r)$ on both sides and find that:

$$\frac{x_2}{1+r} - \frac{I_2}{1+r} = s$$

We then substitute this expression into (2.1) and find the budget constraint for both periods as:

$$x_1 + \frac{x_2}{1+r} = I_1 + \frac{I_2}{1+r} \quad \text{Budget constraint for both periods} \tag{2.3}$$

The left side of the equation above is the present value of consumption, where future consumption is discounted by the interest rate r, while the right side is the present value of income. Figure 2.6 illustrates the choice set. If Anna saves all the money she earns in period 1 ($s = I_1$ so that $x_1 = 0$), she can consume $x_2 = I_1(1 + r) + I_2$ in period 2. This is her maximum consumption in that period and thus gives the intercept between the budget line m and the vertical axis.

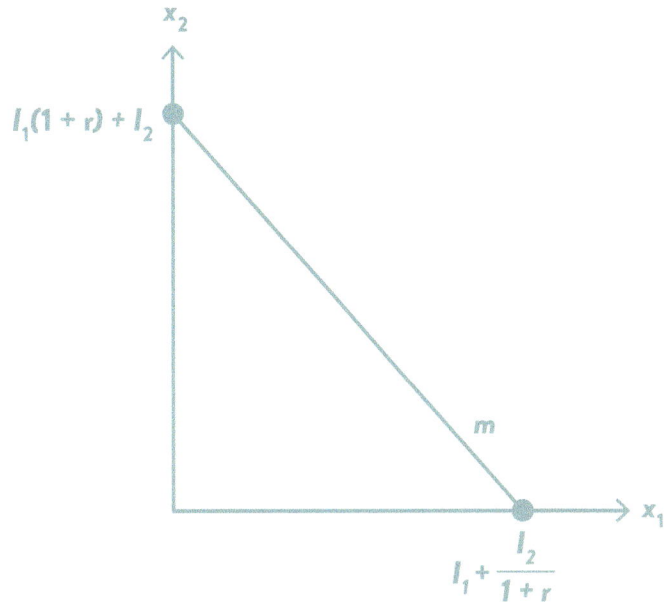

Fig. 2.6 The choice set for consumption in the two periods. *Note* The figure shows the budget constraint in a two-period model, where the consumer has income I_1 in period 1 and I_2 in period 2. She must allocate this income between consumption in period 1, x_1, and consumption in period 2, x_2. If the entire income is saved, the consumer can achieve a maximum consumption in period 2 of $x_2 = I_1(1+r) + I_2$. This is shown as the intercept between the budget line m and the vertical axis. If she borrows the maximum amount, she can consume $x_1 = I_1 + \frac{I_2}{1+r}$ in period 1. This is the point where the budget line crosses the horizontal axis

The maximum she can consume in period 1 is found by setting $x_2 = 0$, which gives $x_1 = I_1 + \frac{I_2}{1+r}$. This is the intercept between the budget line and the horizontal axis.

Anna has decided to work less in the next period, and in Fig. 2.7a this is illustrated $I_1 > I_2$. If she consumes what she earns in each period, that is, $x_1 = I_1$ and $x_2 = I_2$, we get a choice at point a on the budget line m. All choices on the budget line to the left of point a mean she consumer must save, but it is also possible to choose to the right of point a, and to do so she must borrow. Anna wants to have balanced consumption over time, at point b, and to do so she knows she must save, since income in period 2 is not sufficient, since $x_2 > I_2$.

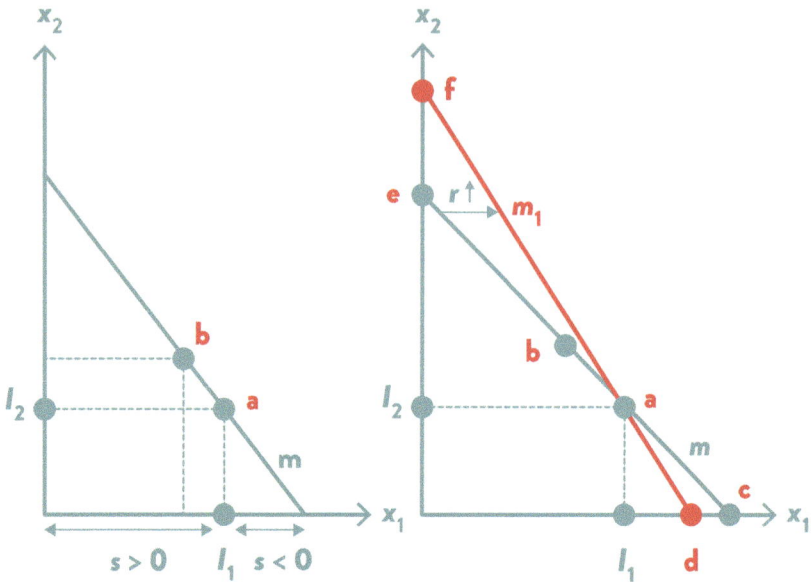

Fig. 2.7 (**a**) Saving and borrowing. (**b**) The effect of a higher interest rate. *Note* Consider first
(**a**). At point *a* on the budget line *m*, consumption equals income in each period, and the consumer
neither saves nor borrows. A choice to the left of *a*, such as point *b*, means the consumer saves,
s > 0, while a choice to the right of point *a* means that the consumer borrows, *s* < 0. (**b**) shows
that a higher interest rate makes the budget line steeper, represented by m_1; it pivots around point *a*,
where saving is zero. A higher interest rate increases the consumption possibilities for a saver: the
maximum period 2 consumption increases from point *e* to *f*. In contrast, it reduces the consumption
possibilities for a borrower: the maximum period 1 consumption goes down from point *c* to *d*.

In Fig. 2.7b, we study the effect of a change in the interest rate, here shown
as an increase. Notice that a change in the interest rate causes the budget line
to pivot around point *a*, where $x_1 = I_1$. Why? Because here *s* = 0, and the
consumer neither has money in a savings account nor borrowed from the bank, so
interest rate changes do not affect consumption possibilities.

For someone like Anna, who has chosen point *b* and thus saved money, the
higher interest rate is good news—it makes her savings more valuable. But for a
person who initially chose a point to the right of *a* and took out a loan, the interest
rate increase is bad news: a higher interest rate means more expensive loans, which
reduces consumption possibilities.

Which point on the budget line will the consumer choose? That depends on
preferences. In Fig. 2.8, we have drawn Anna's indifference curve, U_b, which is
tangent to the budget line at point *b*. With these preferences, Anna chooses to save
s_b. See Math Box 2.1 for a more detailed explanation of the conditions for optimal
choice in a two-period model.

Fig. 2.8 Optimal choice in the two-period model. *Note* The figure shows the optimal choice in the two-period model, given by point b where the indifference curve is tangent to the budget line. Consumption in period 1 is x_1^b while consumption in period 2 is x_2^b. Saving is the income in period 1 that is not consumed, that is, $s_b = I_1 - x_1^b$

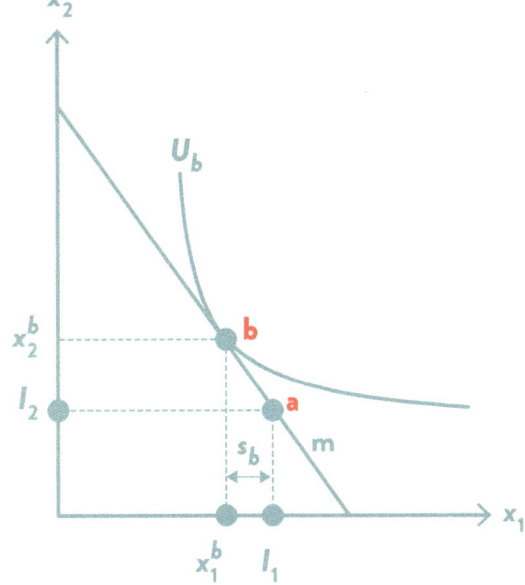

Math Box 2.2: Optimal Choice in the Two-Period Model

To find the optimal choice in a two-period model, we use the same three steps to the optimum as described in Recipe 1 in Chapter 1.

We begin by specifying the utility function we will use, a general Cobb-Douglas function:

$$U = x_2^\alpha x_1^{1-\alpha}$$

Here, α is a measure of how much weight the consumer places on consumption in period 2 relative to consumption in period 1.

Step 1 tells us to use the specific utility function to find the marginal rate of substitution (MRS). From Chapter 1 (see Math Box 1.1), we know that the slope of the indifference curve, the MRS, in this case is given by:

$$MRS = \frac{(1-\alpha)}{\alpha} \frac{x_2}{x_1}$$

Step 2 says that we should use this expression in the condition for optimal consumption, meaning that the MRS must equal the slope of the budget line. To find the slope of the budget line, we start by rewriting (2.3) as follows:

$$x_2 = (I_1 - x_1)(1 + r) + I_2$$

We then differentiate this expression with respect to x_1 and find that:

$$-\frac{\partial x_2}{\partial x_1} = 1 + r \quad \text{Slope of the budget line}$$

We see that the slope is determined by the interest rate: the higher r is, the steeper the budget line. The optimum condition states that:

$$MRS = \frac{(1 - \alpha)\, x_2}{\alpha \quad x_1} = 1 + r$$

The optimal consumption over time is therefore:

$$\frac{x_2}{x_1} = (1 + r)\frac{\alpha}{1 - \alpha}$$

Step 3 asks us to solve the optimal consumption combination with respect to one of the goods.

$$x_2 = \frac{x_1(1 + r)\alpha}{1 - \alpha}$$

We then substitute this expression of x_2 into into the budget constraint (2.3):

$$x_1 + \frac{x_2}{1 + r} = I_1 + \frac{I_2}{1 + r}$$

And find:

$$x_1 + \frac{\alpha x_1}{1 - \alpha} = I_1 + \frac{I_2}{1 + r}$$

We put the expressions on the left side under a common denominator:

$$\frac{x_1(1 - \alpha) + \alpha x_1}{1 - \alpha} = I_1 + \frac{I_2}{1 + r}$$

Which simplifies to:

$$x_1 = (1 - \alpha)\left(I_1 + \frac{I_2}{1 + r}\right) \quad \text{Optimal consumption in period 1}$$

Since saving is given by $s = I_1 - x_1$, we can use the expression for the optimal consumption in period 1 to find the optimal saving as:

$$s = \alpha I_1 - (1 - \alpha)\left(\frac{I_2}{1 + r}\right) \quad \text{Optimal saving}$$

We see that the greater weight the consumer places on consumption in period 2, that is, the larger α is, the more she wants to save. Furthermore, we see that the higher the income in period 1, I_1, relative to period 2, I_2, the more she wants to save. A higher interest rate reduces the present value of income in period 2, and in the same way as a lower income I_2, this contributes to more saving. Note that in the special case where the consumer has no income in period 2, only in period 1, a change in the interest rate does not affect saving.

Finally: We can easily find the optimal consumption in period 2 by substituting the expression for the optimal consumption in period 1 into budget constraint (2.3):

$$x_1 + \frac{x_2}{1 + r} = I_1 + \frac{I_2}{1 + r}$$

$$(1 - \alpha)\left(I_1 + \frac{I_2}{1 + r}\right) + \frac{x_2}{1 + r} = I_1 + \frac{I_2}{1 + r}$$

$$x_2 = \alpha(I_1(1 + r) + I_2) \quad \text{Optimal consumption in period 2}$$

2.7 Summary

In this chapter, we have taken a closer look at what happens when the price of a good or income changes. We have divided the price effect into two steps: the substitution effect and the income effect. Furthermore, we have seen that the effect of an income change depends on whether the goods are normal or inferior. For normal goods, higher income means higher consumption, and lower income means lower consumption, whereas for inferior goods, the relationship is the opposite. This analysis has been useful for understanding the paradoxical situation where a price increase leads to increased demand (Giffen goods) and how the effect of an earmarked gift will partly be offset by a rearrangement of the recipient's own funds (fungibility).

We have also studied saving, which is an important application of consumer theory. The trade-off here is not between pizza and beer, but between consumption now versus consumption in the future. Consumers who expect to earn less in the next period may want to save to smooth consumption over time. However, consumers may also choose to borrow money if they want to consume more than they earn today. The effect of an interest rate change depends on whether the consumer has chosen to save or borrow: For a saver, an interest rate increase is good news, while for a borrower, it is bad news.

P.S. Before we move on to the next chapter, I want to say a few words about Brian. He quit the School of Economics after his first year and went to the School of Engineering where he wanted to pursue his interest in math and technology. At least, that was the plan. Brian's mother is quite worried about her son at this point, and maybe you as the reader also feel some concern. But we will meet Brian again later in the book, and by then some time will have passed... so we'll see how it turned out!

2.8   Key Terms

Income effect: The change in demand caused by a change in purchasing power.
Substitution effect: The change in demand caused by a change in relative prices, holding the utility level constant.
Normal goods: Goods that people consume more of when income increases, and less of when income decreases.
Income expansion line: Shows how the composition of consumption changes as income changes.
Inferior goods: Goods that people consume less of when income rises, and more of when income falls.
Giffen good: A good whose demand goes up if the price goes up, and down if the price goes down.
Fungibility: A rearrangement of one's own funds after receiving an earmarked gift.

2.9   Multiple-Choice Exercises

2.1: Substitution Effect
Consider the change in the budget line from m_1 to m_2. As a result of this change, the consumer changes their choice from point *a* to point *b*. Which of the points 1, 2, 3, or 4 can represent the substitution effect?

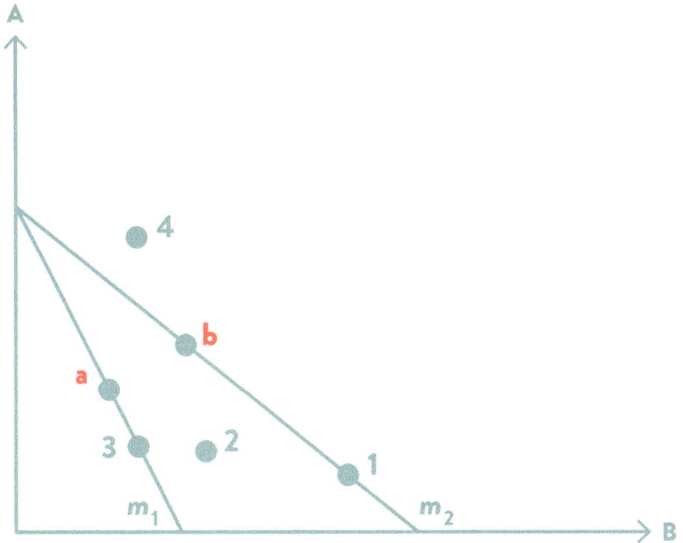

A. 1

B. 2

C. 3

D. 4

2.2: Normal or Inferior Goods?

What is true about the goods shown in the figure above (that is, for the exercise 2.1)?

A. Both A and B are normal goods

B. Both A and B are inferior goods

C. A is a normal good, while B is an inferior good

D. A is an inferior good, while B is a normal good

2.3: Giffen Good

A price decrease on a Giffen good will lead to:

A. Lower consumption of this good

B. Negative income effect for this good

C. Positive substitution effect for this good

D. All of the above

2.4: Saving

In a two-period model with income only in period 1, a consumer with Cobb-Douglas preferences will respond to an interest rate increase by:

A. Reducing consumption in both periods
B. Increasing consumption in both periods
C. Increasing consumption only in period 1
D. Increasing consumption only in period 2

Solutions: 2.1 B; 2.2 C; 2.3 D; 2.4 D

Consumers at Work

3

Why do top managers get paid so much? And which manager ends up with the biggest paycheck—the one who lives for work or the one who values free time a lot?

Anna's boss offers her a higher salary, hoping she will work more, but she declines the offer saying she's happy with her current working hours. Why doesn't Anna want to work more even though the salary increases?

3.1   Introduction

Anna has a part-time job at the grocery shop, a short bike ride from the School of Economics. The job gives her an income she can spend on things she enjoys, but it also takes up time.

"How many hours a week should I work?" she asks herself.

Anna heading to work

K. Bjorvatn, *Microeconomics Made Simple*, Classroom Companion: Economics,
https://doi.org/10.1007/978-3-032-06354-0_3

We analyse labour supply as the trade-off between the joy of leisure and the joy of consumption. More leisure means fewer hours at work, and therefore less money to spend. The optimal choice is determined by prices (which in this case is the real wage, that is, the wage divided by the price of goods) and preferences, just like in the previous two chapters. Consumer theory can also be used to study career choice. A managerial position has become available at the grocery shop, a job that would give Anna bigger responsibilities but little free time. What must the wage be for her to be willing to take on that role?

Sometimes we receive money without working for it, for example a gift from a generous grandfather or unemployment benefits from the government. Toward the end of the chapter, we will examine how such non-labour income affects labour supply.

3.2   How Many Hours Do You Want to Work?

Anna works part time at the grocery shop alongside her studies. The job offers a decent wage and, importantly, the freedom to decide how much she wants to work, which is a clear advantage. A key decision for Anna is how many hours a week she should work. Perhaps you recognise this dilemma?

We will use economic theory to study Anna's choice, starting with the choice set determined by income and prices. To begin with, we assume the job is her only source of income, which is then given by is the hourly wage w multiplied by the number of hours worked J. The money she earns is used to consume M units of goods at a price p per unit (we make no distinction between different kinds of goods, we simply refer to it as *consumption*), so the budget constraint is:

$$wJ = pM \quad \text{Budget constraint} \tag{3.1}$$

Let T be the time she has available after other necessary activities are taken care of (such as sleeping, eating, preparing for the next microeconomics lecture, etc.). This available time can be used for work J and leisure, that is her free time (F). The time budget is therefore:

$$T = J + F \quad \text{Time budget} \tag{3.2}$$

Which implies that time on the job is:

$$J = T - F \quad \text{Working time} \tag{3.3}$$

We can substitute this expression for working time into the budget constraint and get:

$$w(T - F) = pM \quad \text{Budget constraint} \tag{3.3}$$

The choice set can be illustrated using a budget line m, as shown in Fig. 3.1.

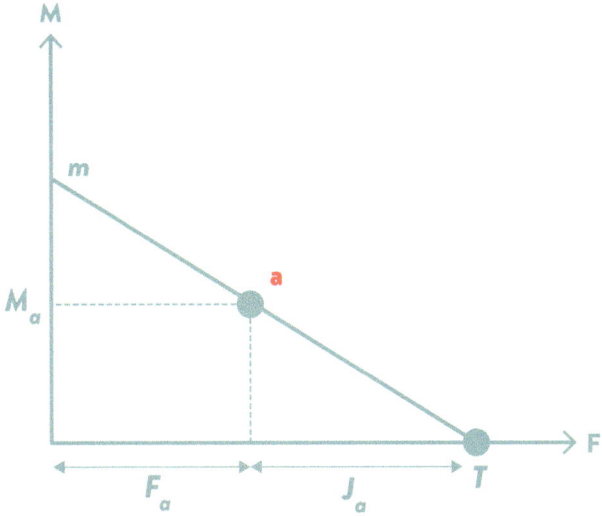

Fig. 3.1 The Worker's Budget Constraint. *Note* The figure shows leisure (F) on the horizontal axis and consumption (M) on the vertical axis. The total available time is marked by point T on the horizontal axis, and labour supply is measured from right to left. The more she works, the higher her income and consumption, as shown by the budget line m. If she works J_a hours, she earns M_a, represented by point a on the budget line

We can express the budget line mathematically by rewriting the budget constraint as:

$$M = \frac{w}{p}(T - F) \quad \text{Budget line}$$

The slope of the budget line is found by differentiating with respect to F (and moving the minus sign so that the right-hand side becomes positive):

$$-\frac{\partial M}{\partial F} = \frac{w}{p} \quad \text{Slope of the budget line}$$

We see that the slope is given by w/p, which we also refer to as the real wage—that is, the purchasing power of the wage when considering the price of consumption goods. A higher wage thus means a steeper budget line. Note that w can be interpreted as the price of leisure: it is the wage one foregoes by not being at work, what we call an opportunity cost or a shadow price.

The number of working hours J is measured from right to left in the figure; this is the time not spent on the sofa (or wherever you spend your free time) but on the job. The maximum consumption is found where the budget line crosses the vertical axis: here, she spends all her time working, so $F = 0, J = T$.

Anna's choice of working hours is determined by the trade-off between leisure and consumption. The best choice is the point where the indifference curve U is

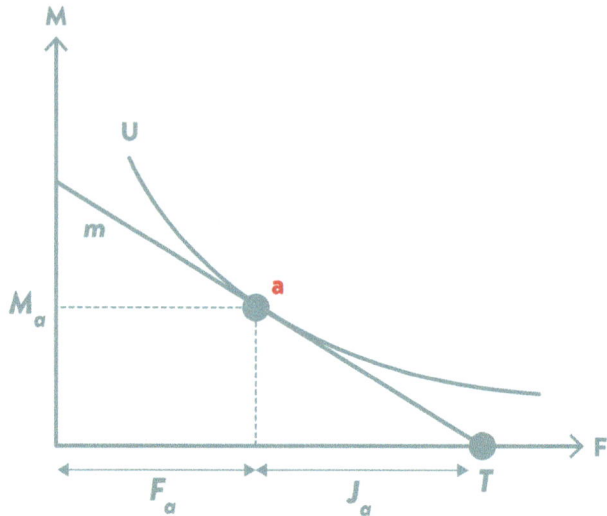

Fig. 3.2 Consumer's Choice of Working Hours. *Note* The figure shows leisure (F) on the horizontal axis and consumption (M) on the vertical axis. Point T indicates where all available time is spent on leisure. Working time (J) is measured from right to left. The utility-maximising choice of working hours is found where the indifference curve is tangent to the budget line m, here at point a, meaning the consumer chooses leisure F_a and thus working hours $J_a = T - F_a$

tangent to the budget line m, shown as point a in Fig. 3.2. Here, Anna chooses to have F_a hours of leisure and work J_a hours. This choice is formally derived in Math Box 3.1, based on a specific utility function.

Math Box 3.1: Utility Maximisation and Labour Supply
Assume the general CD-utility function:

$$U = M^\alpha F^{1-\alpha}$$

We use the same method to find the path to the optimum as in Chapter 1. Step 1 asks us to find the MRS based on the specific utility function. We know that the MRS equals the ratio of the marginal utilities of the two goods. The marginal utility of your free time, or leisure, is:

$$MU_F = \frac{\partial U}{\partial F} = (1 - \alpha)M^\alpha F^{-\alpha} \quad \text{Marginal utility of leisure}$$

Similarly, we can find the marginal utility of consumption as:

$$MU_M = \frac{\partial U}{\partial M} = \alpha M^{\alpha-1} F^{1-\alpha} \quad \text{Marginal utility of consumption}$$

The slope of the indifference curve is therefore:

$$MRS = \frac{MU_F}{MU_M} = \frac{(1-\alpha)M^\alpha F^{-\alpha}}{\alpha M^{\alpha-1}F^{1-\alpha}} = \frac{(1-\alpha)}{\alpha}\frac{M}{F}$$

Step 2 then asks us to find the optimal consumption combination. We start with the condition for optimal choice, which in this case means that the marginal rate of substitution must equal the real wage w/p:

$$MRS = \frac{(1-\alpha)}{\alpha}\frac{M}{F} = \frac{w}{p} \quad \text{Optimal consumption rule}$$

From this, we can find the optimal consumption combination as:

$$\frac{M}{F} = \frac{\alpha}{(1-\alpha)}\frac{w}{p} \quad \text{Optimal consumption combination}$$

We're now done with Step 2.

Step 3 asks us to rewrite the optimal consumption combination so that one of the goods stands alone on the left-hand side. Here, we choose to do this for consumption and find that:

$$M = \frac{F\alpha}{(1-\alpha)}\frac{w}{p}$$

We then substitute the expression for M into the budget constraint $w(T-F) = pM$, and get:

$$w(T-F) = p\left(\frac{F\alpha}{(1-\alpha)}\frac{w}{p}\right)$$

This simplifies to:

$$F = (1-\alpha)T \quad \text{Optimal choice of leisure}$$

Inserting this in the time budget, expressed as $F = T-J$, we can find the optimal choice of time at the job as:

$$J = \alpha T \quad \text{Optimal choice of labour supply}$$

The consumer spends a share α of her available time T working, and the rest on leisure. Note that a person with a balanced Cobb-Douglas utility function,

$\alpha = 0.5$, divides her available time equally between work and leisure. To find consumption in the optimal allocation, we start from the budget constraint:

$$wJ = pM$$

Substituting the optimal choice of labour supply $J = \alpha T$, we find:

$$w\alpha T = pM$$

Consumption can then be written as:

$$M = \frac{w\alpha T}{p} \quad \text{Optimal consumption choice}$$

3.3   What Job Do You Want?

The job at the grocery shop gives Anna an income, but also time left over for studies and leisure activities. She feels she has found a good work-life balance. Now, her boss has asked whether she would consider a managerial position—something that would, of course, require more hours at work. "We can discuss the pay," the boss says, "but the working hours are fixed: J_b hours per week."

At present, Anna works J_a hours per week, giving her an income of M_a, marked as point a on the budget line. She wonders whether she should apply for the managerial position—and, if so, what hourly wage she would be satisfied with. Fig. 3.3 illustrates the situation.

Anna's utility is represented by the indifference curve U. Point b shows how much she must earn to be just as happy working J_b hours in the managerial job as she is with her current job. The minimum acceptable wage for the managerial position is found by drawing the budget line m_2 through point b. We call this the *reservation wage*: she will not be willing to take the managerial job for a wage lower than this. The income at this point, M_b is, as you can see, much higher than her current income M_a.

This higher income is due to two factors: more hours and a higher hourly wage. The effect of more hours is seen at point c, where we follow the original budget line up to the new number of working hours, J_b, resulting in an income increase from M_a to M_c. The movement from M_c to M_b is due to the higher hourly wage.

The high salaries of managers are often debated and may be due to many things—such as significant responsibility or specialised skills—but they can also be understood within this model: managers work long hours, and the wage must be high to compensate for the loss of leisure (the marginal utility of leisure becomes very high when you have little of it, as shown by a steep indifference curve).

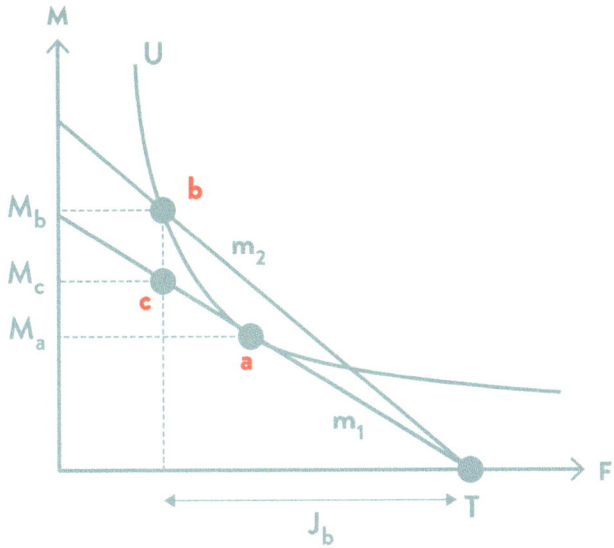

Fig. 3.3 Should She Apply for the Managerial Job? *Note* Her current job at the grocery shop gives Anna a utility level of *U*. The managerial job requires her to work J_b hours, and to be just as happy with that as with her current job, she must earn an income of M_b, given by point *b* on the indifference curve. This implies a higher wage, as shown by the steeper budget line m_2. The increase in income from M_a to M_c is due to working more hours in the managerial job, while the increase from M_c to M_b reflects the higher hourly wage needed to compensate for less leisure

An interesting point is that if people don't value their free time much, salaries will be lower than if they really care about having time off. If you look at Fig. 3.3, Anna would have had a flatter indifference curve if she placed less weight on leisure, and she would therefore have accepted the managerial job even with a lower wage. So, placing less value on leisure leads to a lower reservation wage. And since employers generally don't pay more than they must, the theory predicts that managerial salaries will be lower in a world full of workaholics than in a world full of leisure-lovers.

3.4   Wages and Labour Supply

Anna has decided to turn down the managerial position and continues in her old job with flexible working hours. It's a busy time at the grocery shop, and her boss has offered her a higher hourly wage, hoping it will tempt her to work more. But will it?

Let's study Anna's decision using Fig. 3.4. If the hourly wage increases, the budget line rotates from m_1 to m_2. As discussed in Chapter 2, we can split the price effect into two components: the substitution effect, which is the movement from point *a* to point *b*, and the income effect from point *b* to point *c*. You remember

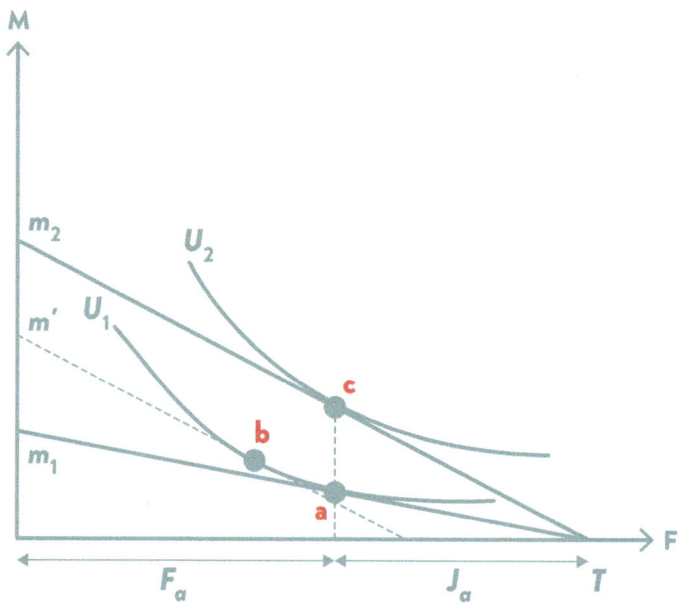

Fig. 3.4 Wages and Labour Supply: Substitution Effect and Income Effect. *Note* The figure shows how an increase in the real wage, which causes the budget line to rotate from m_1 to m_2, affects labour supply (J) and consumption (M). The effect is divided into a substitution effect (from a to b) and an income effect (from b to c). To find the substitution effect, we use the helper line m', which is a parallel shift of the new budget line down to the original indifference curve

how we found the substitution effect, right? You take the new budget line and shift it so that it touches the original indifference curve—see the dashed helper line in the figure. For a refresher on this theory, see Fig. 2.1 and Math Box 2.1.

The substitution effect is unambiguous: a higher wage makes leisure relatively more expensive, in the sense that instead of relaxing in her flat, Anna could have been working at a higher hourly wage at the grocery shop. This pushes her toward wanting to work more—that is, less leisure.

On the other hand, the higher wage expands the choice set, and if leisure and consumption are both normal goods, Anna will use the expanded opportunities to both consume more and enjoy more leisure. And more leisure means fewer working hours. This is the movement from point b to point c.

In Fig. 3.4, we see that these two effects on leisure cancel each other out, so the wage increase does not affect labour supply. This is consistent with the Cobb-Douglas utility function used in Math Box 3.1, where we found that the consumer spends a fixed share of available time working, $J = \alpha T$, regardless of the wage.

Note that the exact cancellation of the substitution and income effects is specific to Cobb-Douglas preferences—other types of preferences, which may produce flatter or more L-shaped indifference curves, will imply different trade-offs.

But the general insight remains: a price change, such as a wage increase, has both a substitution effect and an income effect. And it's important to consider both when predicting the consequences of a price change.

Anna's boss assumed that the higher wage would tempt her to work more, focusing only on the substitution effect. He failed to consider that with the higher wage; Anna could afford to work less and still consume more. He did not understand the income effect of the wage increase.

3.5   Manna from Heaven

In her second year of studies, Anna received a pleasant phone call from her grandfather, Conrad, whom we will get to know better later in the book. He had decided to give some money to his granddaughter and thought that a fixed monthly transfer might be a good way to do it.

"Manna from heaven," Anna thinks when she hears the good news.

Anna receives money from her grandfather

Fig. 3.5 shows how the cash transfer S from her grandfather affects the budget line, which shifts from m_1 to m_2. Note that even if Anna doesn't work at all, that is, chooses $F = T$—she can still afford some consumption thanks to the cash transfer, given by $M = S$. And if she chooses to work, her consumption possibilities expand along the new budget line m_2.

Without support from her grandfather, Anna chooses point a, working J_a hours. But with the cash support, she chooses point b, working J_b hours. We see that the cash transfer has led to a lower labour supply. This makes sense: if leisure is a

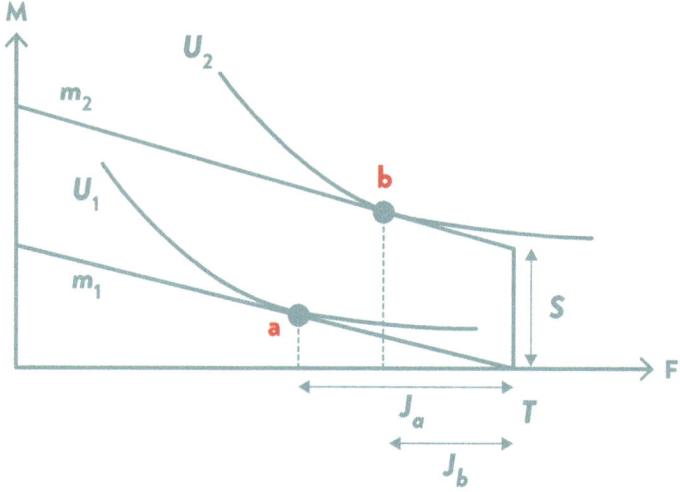

Fig. 3.5 Cash Support and Labour Supply. *Note* The non-labour income S causes a parallel shift in the budget line from m_1 to m_2. Note that if the consumer does not work at all, that is, chooses $F = T$, she can still have positive consumption financed by the cash support. Point a is the preferred choice without the cash support, while point b shows the preferred choice with the support. The cash support is a pure income effect, and if leisure is a normal good, this leads to a lower labour supply: $J_b < J_a$

normal good, she will want more of it when her income rises. And more leisure necessarily means less work.

The second story I want to tell is about Brian. He dropped out of the School of Economics and has applied to the School of Engineering and is now waiting for a response. In the meantime, he isn't quite sure what to do. He's considering getting a job, but the alternative is to receive unemployment benefits from the government. The welfare state ensures that no one goes hungry. It provides money to those who don't have a job or who earn too little to meet their basic needs.

Assume that the authorities have defined a minimum level of consumption, M_{min}, which we can think of as a poverty line. Those who earn less than this can receive support to bring them up to that minimum. Those with incomes above the poverty line, however, do not qualify for support. This is a kind of manna from heaven, but it differs from what Anna received: while the gift from Conrad was unconditional, the state support is conditional on Brian not earning too much.

In Fig. 3.6, we see that the budget constraint has a kink at the poverty line and then flattens out. If Brian earns less than what's required to meet basic needs, he can go receive unemployment benefits. In fact, the less he earns, the larger the support—enough to bring him up to M_{min}.

Brian wonders what to do: work or receive unemployment benefits (he doesn't view going on benefits as something negative—it's like a public sector job that gives him a small income and lots of free time). The choice depends on many

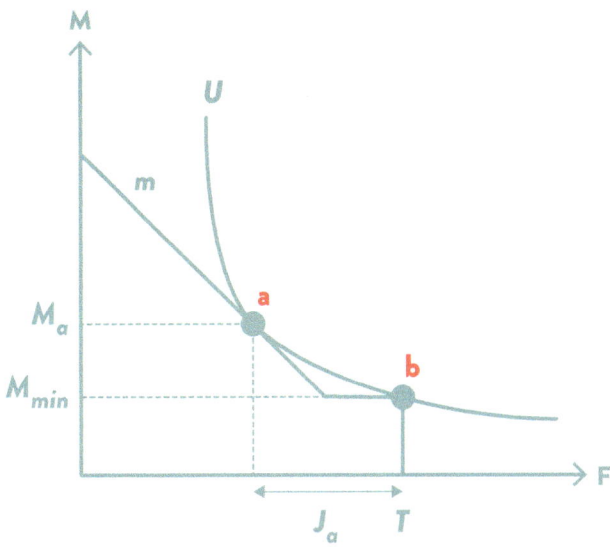

Fig. 3.6 Welfare Benefits and Labour Supply. *Note* The budget line has a kink at M_{min}, the minimum consumption level defined by the authorities. Income below this qualifies for support, and the lower the income, the greater the support. Those without any work receive the most support, where their entire consumption is financed through a transfer from the government. In this figure, the consumer is indifferent between working, point a, and not working, point b

things: what kind of wage he could earn if he works, the size of the benefits, but also his preferences—how he values consumption relative to leisure.

In the figure, we see that if Brian chooses to work, his best choice is a, with working hours J_a. If he doesn't work, he ends up at point b, where all his consumption is paid for by the government. In this case, Brian is indifferent between working and going on benefits. How would a wage increase affect his labour supply? You can explore this question in the multiple-choice exercise 3.4.

How people respond to various support measures is important when designing welfare policy. The goal is to help those in need, while at the same time avoiding a situation where people choose benefits over work. An interesting proposal that has been widely discussed in recent years is the use of a universal basic income—an unconditional cash transfer. What are the consequences of this for willingness to work, compared to today's systems where support is conditional on not earning too much? You can learn more about this by working on Exercise 3.5 in the Workbook.

3.6   Summary

In this chapter, we have examined how consumer theory can be used to analyse choices about work and labour supply. Different jobs offer different combinations of leisure and income, and we have seen that managers earn a high income not only because they work many hours but also because their hourly wage must be high to compensate them for the imbalance between work and leisure.

Many jobs give you some freedom to choose your working hours, and the budget line defines the choice set between leisure and consumption. The optimal choice of working time is found where the indifference curve is tangent to the budget line.

Finally, we studied how non-labour income, whether in the form of a cash transfer or unemployment benefits from the government, affects labour supply.

In this chapter, we also briefly met Anna's grandfather. He is the one who provides her with the monthly money transfer. Conrad, as he is called, owns a paper mill in a small industrial town, and we will get to know him and his businesses better later in the book.

3.7   Key Terms

Opportunity Cost: The value of time or money in its best alternative use. Opportunity cost is also called shadow price.
Reservation Wage: The lowest wage you are willing to work for.

3.8   Multiple-Choice Exercises

3.1: Shift in the Budget Line
Imagine a student with a part-time job who initially faces the budget line m_0, with available time T_0. Now assume that exam period is approaching, which we can interpret as a reduction in the available time for work and leisure. Which of the new budget lines captures this changed situation?

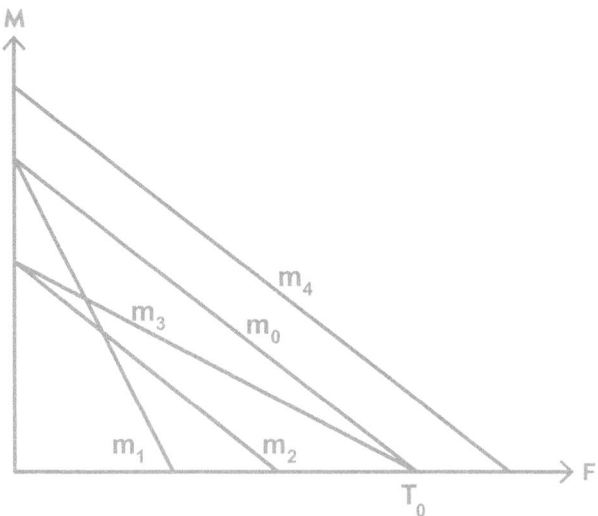

A. m_1
B. m_2
C. m_3
D. m_4

3.2: Preferences for Leisure

A student chooses to work five hours a week out of a total available time of 20 h. Assume the utility function is $U = M^\alpha F^{1-\alpha}$. What is the value of α?

A. 0.15
B. 0.20
C. 0.25
D. 0.30

3.3: Substitution Effect in Labour Supply

Which of the following statements is/are correct?

1. The substitution effect of a higher wage means the consumer chooses to work fewer hours.
2. The income effect of a higher wage means the consumer chooses less leisure.
3. If the wage goes up and the consumer chooses to work more hours, the substitution effect dominates the income effect.
4. If the wage goes down and the consumer chooses to work more hours, the income effect dominates the substitution effect.

A. 1, 2 and 4
B. 1 and 4

C. Only 3
D. 3 and 4

3.4: Work or Welfare?
Consider Fig. 3.6 depicting a person who is indifferent between working (point a)
and receiving welfare (point b). Assume the person has Cobb-Douglas preferences.
What happens if the wage level (w) increases?

A. He will choose to work J_a hours.
B. He will choose to work more than J_a hours.
C. He will choose work over welfare, but we cannot say how many hours he will
   work.
D. He will still be indifferent between work and welfare.

Solutions: 3.1 B; 3.2 C; 3.3 D; 3.4 A

Behavioural Economics

4

Taxi drivers may wait for customers for hours on a slow day, but choose a short working day when there is a queue of customers: Why?

Brian finds a job, and his mother increases her consumption: How are these connected?

4.1 Introduction

We have studied many choices: Anna's choice between housing and other things, Brian's choice between pizza and beer, and both of our friends' choice between consumption and leisure. The starting point has always been that consumers do the best they can with the money they have and the prices they face in the market. Preferences are stable, and they maximise their utility. In other words: consumers are rational. Moreover, they are solely concerned with their own consumption when making choices: they are selfish. We call this type *homo economicus*, the economic man.

Anna offers a donation to the Salvation Army

© The Author(s), under exclusive license to Springer Nature Switzerland AG 2026   59
K. Bjorvatn, *Microeconomics Made Simple*, Classroom Companion: Economics,
https://doi.org/10.1007/978-3-032-06354-0_4

Rationality and selfishness were long almost universally accepted assumptions in economic theory. Economists left it to psychologists and sociologists to study irrational behaviour, such as inconsistent choices or behaviour driven by concern for others and moral principles. For example, why Anna willingly stops by the Salvation Army's Christmas collection to make a donation, or why these soldiers volunteer so much of their time for good causes. This is behaviour economists for a long time did not bother trying to understand.

But since the early 2000s, economists' willingness to consider alternative motivations and behaviours has changed radically. The starting signal was, in many ways, in 2002 when the Israeli psychologist Daniel Kahneman (1934–2024) received the Nobel Prize in Economics. The field he championed is called behavioural economics.

Theoretical behavioural economics uses standard economic methods but studies how limited rationality and extended preferences lead to choices that conflict with what we would expect from *homo economicus*. Empirical behavioural economics makes extensive use of experiments, typically based on incentivised tasks so the situation feels important and realistic.

Behavioural economics is broad, and the ambition here is neither to provide a comprehensive overview of the field nor a list of its key contributions. For the interested reader, I recommend Kahneman's classic *Thinking, Fast and Slow* from 2011. The book's title also hints at when limited rationality is especially relevant: It is in situations where one thinks fast, meaning in everyday decisions where not too much is at stake. We often think faster and less rationally in the shop or at a party than when applying for a job or choosing where to live.

In this chapter, I give you some glimpses of behavioural economics, and I do so based on the modelling framework we have covered so far in the book, so you can more easily see how this new perspective relates to traditional theory. The chapter begins with saving and how temptations can cause us to deviate from our long-term saving plans. Then I introduce loss aversion and show how this can affect labour supply. Finally, we will analyse how Brian's mother thinks when deciding how much to give to her son, based on altruistic preferences.

4.2   Temptations and Saving

Anna sees that she will have less time for work next semester and therefore wishes to save to smooth her consumption over time. We studied her saving decision in Sect. 2.6. Anna is quite a disciplined type and will probably manage to follow through on her saving plan. But for many of us, it is challenging to keep up good intentions. Just think of all the broken New Year's resolutions! Did you know that the second Friday of the new year is called *Quitters Day?*

Imagine being presented with two different scenarios. In both cases, you're deciding between getting a smaller amount sooner or a larger amount later. Opting for the larger, delayed amount is essentially a way of saving: rather than taking the money now, you can place it in the bank and earn an interest, allowing for a bigger payout in the future.

Consider the following two scenarios. In each case, what would you choose?

Scenario1: Future payments
100 euros in one year
150 euros in one year and one month

Scenario 2: Present payments
100 euros today
150 euros in one month

Note that in both scenarios, there is one month between the earlier and the later payment. So, in principle, the two situations are the same. If you are patient and choose the 150 euros in the future, then logically you should also choose the same for present payments. And if you are a more impatient soul, you should choose the 100 euros in both scenarios.

But what we often see in experiments like this (where real money is usually at stake) is that many are patient in the long term (Scenario 1) and choose 150 euros, but impatient in the short term (Scenario 2), and then choose 100 euros.

Economists call this *time-inconsistent preferences*. It is a violation of the requirement for logically consistent choices, which is a fundamental assumption of standard economic theory. A behavioural economic explanation for this phenomenon is that people fall for temptations. Deep down, we want to save, but when the choice is imminent, the temptation to take the smaller amount immediately becomes too strong.

We can think of this as having two personalities: your patient self (version A) and your impatient self (version B). In Fig. 4.1, we show the choices for both personalities. For simplicity, assume you have income only in period 1, I_1. Consumption tomorrow therefore requires saving. The preferred choice of your version A is at point a, where the indifference curve U_A is tangent to the budget line m. We see that saving is given by s_a.

That was the plan, and it is easy to be patient when saving is far off in the future. But here and now there are many temptations, and your version B choice might instead be at point b, where saving is much less than originally planned!

Behavioural economics focuses on this type of decision bias and also explores ways to help us make better choices, in line with our long-term preferences. This applies to many areas—not just saving—but where the common factor is a short-term cost to achieve a long-term gain: for example, exercising more, eating healthier, drinking less … and perhaps even starting to prepare now for your microeconomics exam, even though it's still a long way off!

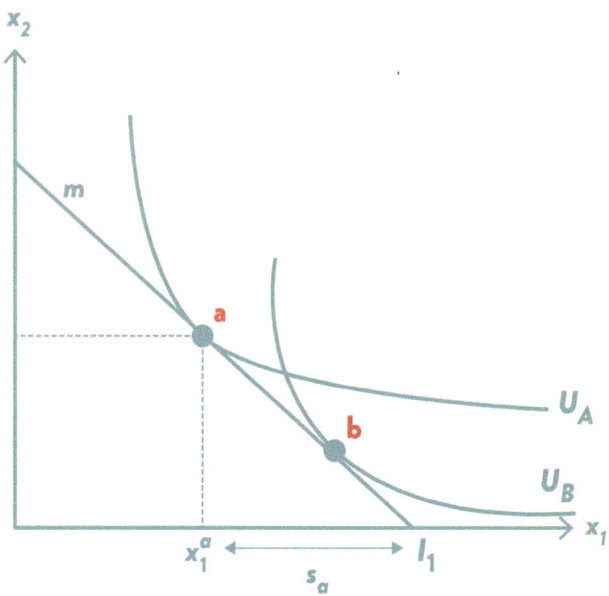

Fig. 4.1 Same person, but two different personalities. *Note* The figure shows the budget constraint in a two-period model with income only in the first period, where the patient version has the indifference curve U_A and chooses point a with saving s_a, and the impatient version has U_B and chooses point b, with less saving

Brian at the reading room: staying up too late the night before, again…

When it comes to temptations and saving, one approach has gained particular attention, building directly on insights from behavioural economics: the Save More Tomorrow programme. This saving plan was developed by the 2017 Nobel laureate in economics, Richard Thaler (born 1945), who is also known for his book *Nudge* from 2008 (which comes highly recommended!). The Save More Tomorrow plan involves making decisions about saving, but the savings agreement only takes effect after a year, thereby exploiting the fact that people are more inclined to save from future income than present income. In the US, several companies have adopted the Save More Tomorrow plan for their employees, helping them save more effectively.

Entering into a savings agreement with your employer can be effective, but typically, you must rely on yourself—and that can be challenging. Imagine again a situation where you only have income in period 1, so you need to save to consume in period 2. As shown in Fig. 4.2a, version A would choose point a, with saving s_a, while Fig. 4.2b shows that version B would choose point b, with less saving. Sometimes, you can be quite disappointed in yourself, and Fig. 4.2a shows that point b clearly gives your patient self a lower utility than point a, $U_A^b < U_A^a$.

A clever trick to avoid falling for the temptation to spend too much money today (from your future impatient self) is for you—that is, your patient self—to open a savings account where the money is locked in, and where you must pay a penalty fee g if you choose to withdraw money from the account in period 1. See Fig. 4.2b.

You have saved an amount s_a, which allows you to consume x_1^a in period 1. To consume more than this today, you must withdraw money from the savings account. If withdrawing money has no cost, you follow the original budget line m, and we see that version B then prefers point b.

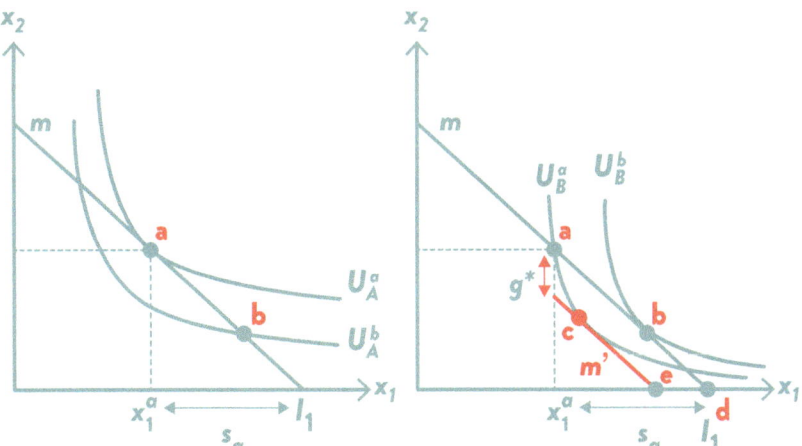

Fig. 4.2 (**a**) Your patient self (A) and impatient self (B). (**b**) A penalty to keep you committed. *Note* (**a**) shows how the patient version A experiences lower utility at point b, which is the preferred choice of the impatient version B. (**b**) illustrates how a penalty fee g for withdrawing money from the savings account shifts the budget line downward. Your impatient self B, facing such a penalty, would choose point c, and we see that the fee in this case makes version B indifferent between using the savings and leaving them untouched. Such a penalty (or higher) would protect your savings from your impatient self

But what if we introduce a fee g for withdrawing money from the savings account? For simplicity, assume the fee is fixed regardless of the amount withdrawn. In this case, the budget line is no longer m, but m'.

Why m'? Imagine you withdraw all your savings to consume everything today. This is the maximum consumption today, originally at point d, but with the fee it is now at point e. If you withdraw from the savings, the budget line lies below the original, by a fixed amount equal to the penalty g.

Version B's optimal choice on m' is at point c, which gives the same utility as point a, U_B^a. In other words, with this penalty, version B is indifferent between not dipping into the savings and smashing the piggy bank, so that a penalty g^* or higher would therefore guarantee that the savings remain untouched. Such a penalty—which you have imposed on yourself (!)—could therefore protect your savings from future temptations.[1]

4.3   Loss Aversion and Labour Supply

Behavioural economic research shows that people tend to perceive changes relative to a reference point, and that the pain of losing is much greater than the pleasure of gaining—about twice as strong, in fact. This is known as *loss aversion*.

Figure 4.3 provides the classic illustration of loss aversion, taken from one of the earliest and most influential contributions to behavioural economics—a 1979 article by Daniel Kahneman and his long-time collaborator, Israeli psychologist Amos Tversky (1937–1996). It depicts a scenario where a person starts out at the origin, then receives 100 euros. According to the utility function, this gain yields a utility of 1. Now imagine a similar thought experiment, but this time the person *loses* 100 euros. The utility function shows a drop in utility of 2. In other words, the pain of losing 100 euros is far greater than the pleasure of gaining the same amount.

One of the most famous studies on loss aversion involves taxi drivers in New York.[2] It is common for drivers to set themselves a daily earnings target, say, twice the cost of renting the car (for those who don't own one). On slow days, they must work longer hours to reach this target, whereas on busy days, they can clock off earlier. In other words: more hours when the hourly income is low, fewer hours when it's high. The income target acts as a *reference point*, and income below this level results in a significant utility loss, something the drivers try to avoid by working long hours.

---

[1] Ashraf, Karlan, and Yin (2006), in their paper *Tying Odysseus to the Mast: Evidence from Commitment Savings Products in the Philippines*, published in the *Quarterly Journal of Economics* (pp. 635–672), provide evidence for such savings strategies. The idea has even been commercialised—see stickk.com: How much are you willing to pay to avoid failing your diet goal?

[2] Colin Camerer, Linda Babcock, George Loewenstein, Richard Thaler (1997). «Labor supply of New York cabdrivers: One day at a time». *Quarterly Journal of Economics*: 407–441.

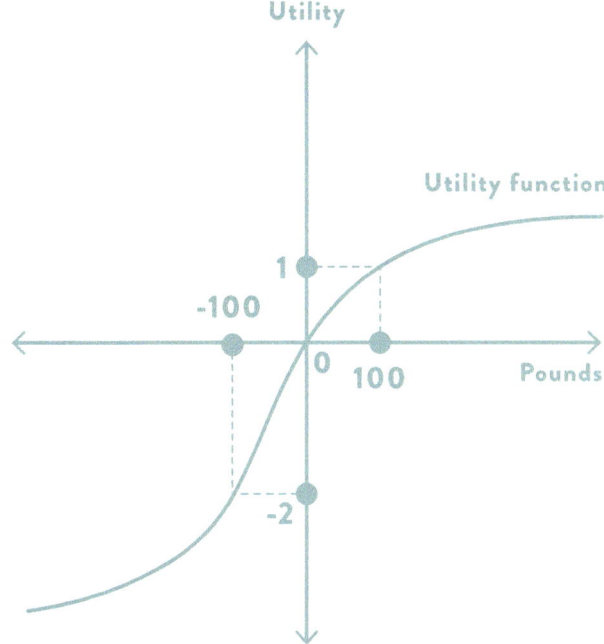

Fig. 4.3 Loss Aversion. *Note* This is the classic illustration of loss aversion. It shows both the positive utility of a gain and the negative utility (that is, the pain) of a loss, where gains are measured to the right from the origin and losses to the left. A gain of 100 euros gives a utility gain of 1, while a loss of 100 euros causes a utility loss of 2

Let's formalise this and compare the labour supply of a taxi driver with such an income target—let's call him Bill—to a driver of the standard *homo economicus* type, Allan. Look at Fig. 4.4, based on the standard framework for analysing labour supply that we're now well familiar with from Chapter 3. We have material consumption, M, on the vertical axis, and leisure, F, on the horizontal axis. Total time is given by T, and working hours are measured from right to left.

Both drivers work six days a week, Monday to Saturday, taking Sundays off. Imagine that customer traffic varies by weekday: early in the week, there are few customers, but demand builds up toward the weekend, when people go out to eat and drink. On Fridays and Saturdays, there is almost no waiting for the next customer, so the budget line is steep: m_c. On Mondays and Tuesdays, hourly earnings are lower due to long waits between customers, so the budget line is flatter: m_b. Wednesdays and Thursdays are more average, represented by budget line m_a.

Assume both Allan and Bill, in midweek, choose point a, with working hours J_n. For Allan, this is about maximising utility, here represented by U_n, and the tangency condition is met at point a. For Bill, it's about hitting a daily income goal, M_0.

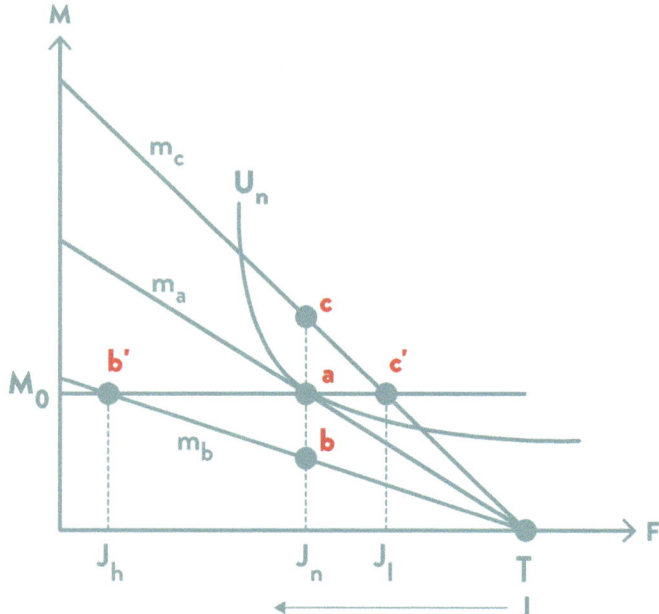

Fig. 4.4 Loss Aversion and Labour Supply. *Note* The figure shows a consumer's choice between consumption (M) and leisure (F), where the budget lines vary from m_c (weekend, high wage), to m_b (early in the week, low wage), and m_a (midweek, average wage). In this case, the consumers are taxi drivers. A driver with Cobb-Douglas preferences chooses to work J_n hours every day. In contrast, a driver with an income target of M_0 adjusts their labour supply: working J_n on a typical day, J_h on a slow day with few customers and a low hourly wage, and only J_l on a busy day with many customers and a high hourly wage

What happens at the end of the week? If Allan's preferences are of the Cobb-Douglas type, we know that his labour supply is unaffected by wage changes (assuming work is only source of income): the substitution effect (higher wage, more work) and income effect (higher wage, more leisure) exactly cancel. So, Allan sticks with J_n, but his income and consumption rise to point c.

What about Bill? Once he hits his daily income target, he goes home—and with higher hourly earnings, he can do that after working just J_l hours, at point c'.

On a slow day, Allan again works J_n, at point b, with lower consumption. Bill, on the other hand, pushes himself. To reach M_0, he needs to work J_h hours on Monday and Tuesday, choosing point b'—and likely collapsing into bed afterwards!

Allan, and other drivers with Cobb-Douglas preferences, prefer a balance between work and leisure, and their labour supply doesn't fluctuate with weekday and pay rate the way it does with drivers with an income target, such as Bill.

Researchers studying New York taxi drivers found that many young and inexperienced drivers behaved like Bill, while the older, more seasoned drivers behaved like Allan. Perhaps not so surprising: spending much of your day waiting for the next customer gets tiresome over time!

4.4 Altruism and Gifts

As we know, Brian's mother sometimes buys him pizza to make sure he has some basic food at home and isn't just living on beer. Admittedly, things didn't go quite as she had planned—but that's another story (covered in Chapter 2).

Here, we're focusing on the gift itself. Brian's mum treats him to pizza because she cares about him. This is what we call *altruism*, or kindness. Of course, in this case, it may not seem all that surprising, since we're talking about a mother and her son—but research shows that people are often willing to share even with strangers, and with no expectation of anything in return—not even recognition or a favour.

The *dictator game* is a classic experiment in behavioural economics used to measure altruism. The setup is incredibly simple: each participant is given a sum of money—say, 100 euros—and asked to decide how much of it they want to give to someone else. The recipient is a stranger, and both sender and recipient remain anonymous. So, the only reason to give away anything at all is the sheer joy of giving.

Homo economicus—the perfectly selfish economic agent—would, of course, give nothing. But in dictator games, many people give half, and on average around one-third of the money is shared.

The dictator game shows that people are altruistic, and altruism breaks one of the core assumptions of standard economic theory. But does that mean everything we've learned in the previous chapters, based on selfish preferences, was wrong or a waste of time? Not at all!

For one thing, altruism isn't a relevant factor in all decisions (how often do you think about generosity when doing your grocery shopping?), and especially not in the kind of market transactions—buying and selling—that are the main focus of this book. For another, we can use the tools of consumer theory to analyse decisions that are driven by altruism.

Let's take an example. Suppose that Audrey (Brian's mum) has a utility function $U = U(A, B)$, where A is Audrey's consumption and B is Brian's. In other words, Audrey cares not just about her own consumption, but also about her son's: she is altruistic.

Brian isn't entirely dependent on his mother. He sometimes earns a bit of money himself and can buy a few basic things. Let I_B be Brian's own income, which he can spend on himself, and let B_A be the gift from Audrey. Brian's consumption can then be written as:

$$B = B_A + I_B \quad \text{Brian's consumption} \tag{4.1}$$

Audrey has an income I_A which she uses on her own consumption (A) and on her son's consumption (B_A). For simplicity, we assume that the price of consumption goods is the same for both and normalise the price to 1. Her budget constraint is therefore given by:

$$I_A = A + B_A \quad \text{Audrey's budget constraint} \tag{4.2}$$

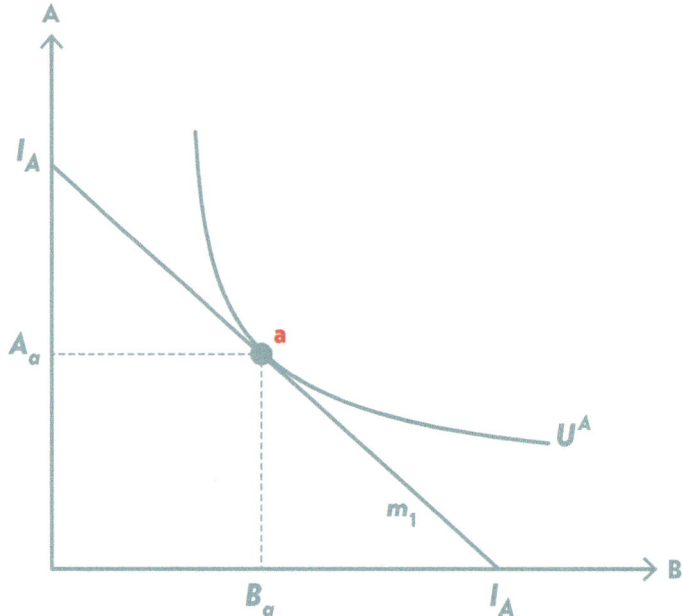

Fig. 4.5 The Altruism Model. *Note* The figure shows the mother's choice between her own consumption (A) and her son's consumption (B) in a situation where the son has no income of his own. The set of feasible combinations of personal and son's consumption is given by the budget line m_1, assuming the price of consumption is the same for both mother and son. The mother's optimal choice is at point a, where her indifference curve U^A is tangent to the m_1 line. We see that she gives B_a to her son

Figure 4.5 illustrates the altruism model, where the opportunity set is defined by the budget line m_1. We assume that Brian initially has no income ($I_B = 0$), meaning he is entirely dependent on his mother's gift for his consumption.

Audrey's optimal choice is given by point a, where her indifference curve U^A is tangent to the budget line m_1. At this point, Audrey consumes A_a herself and gives B_a to her son.

Audrey cares about her son

But what happens if Brian is lucky enough to find a part-time job? Well, this shifts the opportunity set outwards, as shown in Fig. 4.6. Notice that the opportunity set now has a kink at $A = I_A$: If Audrey spends all her money on herself, Brian will still have a positive level of consumption, given by $B = I_B$, based on his own income. If, on the other hand, she gives away all her money ($A = 0$), Brian's consumption will be $B = I_A + I_B$. This defines the point where the budget line intersects the horizontal axis.

There are two interesting points to note from this figure related to Brian's earning money. First, it leads to higher consumption for Audrey, illustrated in the figure as the movement from point a to point b along the budget line m_1. She reduces her contribution to her son, from B_a to B_b, and thereby increases her own consumption from A_a to A_b. The logic is simple: when Brian can at least partly support himself, this frees up resources for Audrey. Brian's wage thus creates a positive income effect for his mother!

Second, Brian's consumption increases by less than you might expect. This is clearly shown in the figure, where his consumption rises from B_a to B_c, and this increase is less than his income, I_B. This phenomenon is known from Chapter 2 as fungibility. Here, the fungibility is based on Audrey re-thinking her spending when Brian gets a job.

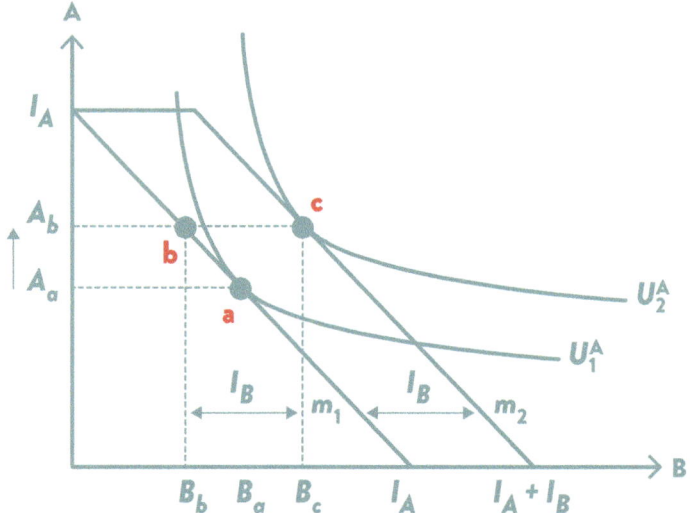

Fig. 4.6 Audrey's response to Brian finding a job. *Note* The figure shows the mother's choice between her own consumption (A) and her son's consumption (B) in a situation where her son now has an income, I_B. This shifts the opportunity set outward, from m_1 to m_2. The new optimum is at point c, where Audrey's indifference curve U_2^A is tangent to the m_2 line. Her son's income leads her to reallocate spending—she gives less to him (from B_a to B_b) and keeps more for herself (from A_a to A_b). So, the increase in her son's income results in higher consumption for Audrey. Brian's consumption also rises, but by less than his income. His total consumption is $B_c = B_b + I_B$, that is, the sum of what he receives from his mother and what he buys with his own money

4.5   Summary

In this chapter, we have seen how key topics in behavioural economics—such as temptation, loss aversion, and altruism—can be analysed using standard microeconomic tools, and how choices are influenced by alternative assumptions about behaviour. For example, we have seen that temptation can cause deviations from a long-term savings plan, and that loss aversion can affect labour supply. We have also studied altruism, where a loving mother makes a gift balancing a concern for her own consumption and that of her son.

Behavioural economics has evolved from being a curiosity in economics—something most economists were somewhat sceptical about—to becoming an integrated part of the field with significant research interest. Large-scale economic experiments are conducted to learn more about how we think and behave in various situations, often using students as test subjects. This research provides new insights that challenge standard models and show when they are relevant and when they may need some adjustment. In this way, science progresses and economics becomes even richer.

4.6 Key Terms

Time inconsistent preferences: When a person is patient in the long run but impatient in the short run.
Loss aversion: The tendency to weigh the pain of a loss much more heavily than the pleasure of an equivalent gain.
Dictator game: A game that measures altruism, where participants are given a sum of money and must decide how much they want to share with a stranger.

4.7 Multiple-Choice Exercises

4.1: Saving and Temptations
Based on Fig. 4.2b, which of the following statements is/are correct?

A. A fee for withdrawing savings can be good for version A, but never for version B
B. The less weight version B places on consumption in period 1, the larger fee version A must choose to protect the savings
C. A lower fee than $g*$ would cause version B to choose point a
D. All of the above

4.2: Loss Aversion
Loss aversion means that people:

A. Dislike losing
B. Dislike losing more than they like winning
C. Dislike losing as much as they like winning
D. Like losing

4.3: Audrey's Altruism
Based on Fig. 4.5. What do you think would happen if altruistic Audrey cared a little less about her son?

A. She would still choose point a, since that is where the indifference curve touches the budget line
B. The budget line would become flatter, and Audrey would therefore spend less money on herself
C. The indifference curve would change, with the optimal choice higher up on the budget line
D. The budget line would become steeper, and Audrey would therefore spend more money on herself

4.4: The Opportunity Set

When Brian in Fig. 4.6 earns his own money enabling him to buy I_B himself, there is a kink in the budget line m_2 because:

A. The mother cannot use the son's income for her own consumption
B. The mother is altruistic, and she will not spend more money on herself
C. There is a minimum amount of support that the mother will always offer her son
D. All of the above

Solutions: 4.1 A; 4.2 B; 4.3. C; 4.4 A

Part II
Producer Theory

In this part of the book, we are introduced to Conrad's paper mill. The main product is A4 paper, and Conrad is wondering how the factory should combine labour and machines to keep production costs as low as possible. Or perhaps he should opt for the new robot technology?

He's also considering expanding production and gets help from Anna, who has just completed her studies at the School of Economics, to calculate the costs and profitability of such an expansion. We also meet Brian again, who has finished his degree at the School of Engineering.

We use the paper mill to study the choices faced by producers in a situation where they take market prices as given—both the price of the product and the prices of the inputs. That is, decision-making in a perfectly competitive market.

In Chapter 5, we analyse the choice of input combination for a given level of output. How should labour and machines be combined to keep costs as low as possible? Or what about fully automating production using new robot technology?

In Chapter 6, we examine what it takes to increase production, and what it would cost in the short and long run to do so—where short run means that capital is fixed. As we'll see, the time frame is crucial to understanding the costs of expanding production.

In Chapter 7, we focus on profitability. In the short run, profitability is about covering variable costs, while in the long run, firms must also ensure that the capital owners are satisfied: they must get a return on their investment that is at least comparable to what they could alternatively have made by placing their money elsewhere—such as in the bank.

Labour and Capital

5

About Conrad and his paper mill: Should he let robots do the job—or perhaps move production to China?

And about Brian, who insists the factory should always have the most modern production equipment, and Anna, who's doing her best to keep costs under control.

5.1   Introduction

Anna's grandfather, Conrad, owns a paper mill in a small industrial town. The factory produces one billion A4 sheets each year. Conrad is not the only paper producer on the market—but we'll come back to that later.

Conrad in front of the paper mill founded by his father,
which he hopes Anna will one day take over.

Production relies on a dedicated group of workers and a fairly modern set of machines. Conrad has run the factory for many years—it's a family business he inherited from his father, and he hopes that Anna will one day take it over.

© The Author(s), under exclusive license to Springer Nature Switzerland AG 2026   75
K. Bjorvatn, *Microeconomics Made Simple*, Classroom Companion: Economics,
https://doi.org/10.1007/978-3-032-06354-0_5

Conrad is a capitalist, but also a gentleman who cares about his employees. He's not as young as he used to be, but fortunately, he has a strong team around him to offer advice.

Some years have passed since the first part of the book. Anna has just completed her degree in economics and has moved back to the industrial town to work for her grandfather. And as if rising from the ashes like a phoenix, Brian reappears in our story. He got into the School of Engineering and, against all odds, graduated with flying colours. Anna has kept in touch with him throughout, and when she heard of an open position at the paper mill, she tipped him off. He applied, was offered the job, and accepted immediately—thanks in no small part to Anna... (You'll have to wait until the final chapter to find out where that relationship goes.)

Conrad is wondering whether there are more cost-effective ways of producing paper. And for years, he's been thinking about expanding production and he wants to know what it would cost, and whether it would be profitable.

We'll answer all these questions in this part of the book. But we will start with production technology: how people and machines can be combined to produce a given quantity of output—and how they *should* be combined to minimise costs. New technology makes it possible to automate the production process. Should Conrad invest in robots and reduce the need for labour to a minimum? In Chapter 6, we'll examine how to scale up production and what it costs, and in Chapter 7, whether it pays off.

5.2   Technology

Conrad likes to keep up with the times and knows that there are different ways to produce paper. He's been reading about labour shortages in various industries and is now considering whether the paper mill should shift to a more capital-intensive production process. He turns to Brian, who studied engineering and knows about this sort of thing.

To produce paper, you need people and machines—or, in economic terms, labour and capital. Of course, you also need raw materials—most importantly pulp for paper production—and electricity to keep the machines running. But when we're thinking about *how* to produce the paper, it's primarily about labour and capital. So, in this chapter, we'll focus on these two inputs.

Brian illustrates the options using Fig. 5.1. On the vertical axis, he plots capital input; on the horizontal, labour input. We can think of labour as the number of full-time workers, and capital as the quantity, or quality, of machinery. These are of course simplifications. In practice, there are different types of workers in a company—for example, some work in administration while others are involved in production. Similarly, there are different types of capital equipment, such as tools, machines, and buildings. But for simplicity, we bundle these together and say *a worker is a worker*, and *a spade is a spade*—where the spade here represents capital equipment. Theory is all about making useful simplifications allowing us to study a complex reality in a manageable way.

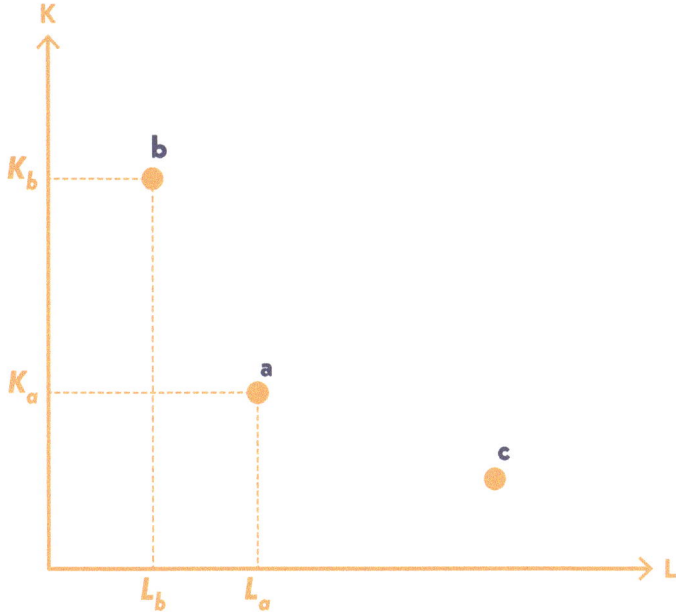

Fig. 5.1 Different combinations of input factors. *Note* The horizontal axis shows the use of labour in production, and the vertical axis shows the use of capital. The three points represent different ways of producing the same quantity: Point *b* reflects a capital-intensive production method, point *c* a labour-intensive one, while production at point *a* relies on a more balanced mix of labour and capital

Brian notes that the factory currently produces one billion sheets of paper per year using L_a workers and K_a units of capital. But there's an alternative: moving to point *b*, where more, or more modern, machines allow the same output to be produced with fewer workers. Brian, who's a big fan of technology, strongly recommends this solution. That said, he points out that there are also other ways to produce paper—for example, at point *c*, where production is more labour-intensive than it is today. That's a solution he does *not* recommend.

Conrad thanks Brian and heads back to his office to reflect. He knows that Brian is fond of machines, perhaps a little too enthusiastic. Conrad, for his part, is a people person and cares about his employees. But his priority is to ensure the profitability of the business. He knows that if he doesn't keep an eye on costs, everyone could lose their job.

He therefore turns to Anna in the finance department. As we know, Anna studied at the School of Economics and knows what things cost, and how to think about profitability.

5.3   Isocost

Conrad tells Anna about his visit to Brian and his proposal to modernise production. He shows her Brian's drawing, and she asks him to take a seat while she pulls out pen and paper.

Anna: Let me show you something I learned in microeconomics. It's called an *isocost line*. It shows different combinations of labour and capital that all result in the same total production cost as we have today.

Conrad: Aha, *iso-cost*. So, same cost. I took some Greek in high school, and if I remember correctly, *isos* means "equal" in Ancient Greek.

Anna shows Conrad Fig. 5.2, where she has added two lines to Brian's earlier Fig. 5.1. These are called isocost lines, and their slope (as we'll soon see) is determined by the factor price ratio, w/r. The isocost line shows all combinations of capital and labour that result in the same cost as production at point a, while isocost line corresponds to the costs at points c and b. Since lies farther out in the diagram than, it reflects a higher cost level.

Today's production (at point a) lies on isocost line C_1. Brian's suggestion of a more capital-intensive production at point b would lead to higher costs—line C_2. And since we know from Brian's figure that points a, b, and c all produce the same quantity, it wouldn't make sense to move from a to b. Nor is the more labour-intensive technology at c any more cost-effective than today's setup. So, Anna at least agrees with Brian that c is not a good solution.

Her advice to Conrad is to stick with the current factor combination, as represented by point a.

Mathematically, the costs can be expressed as:

$$C = wL + rK \quad \text{Costs} \tag{5.1}$$

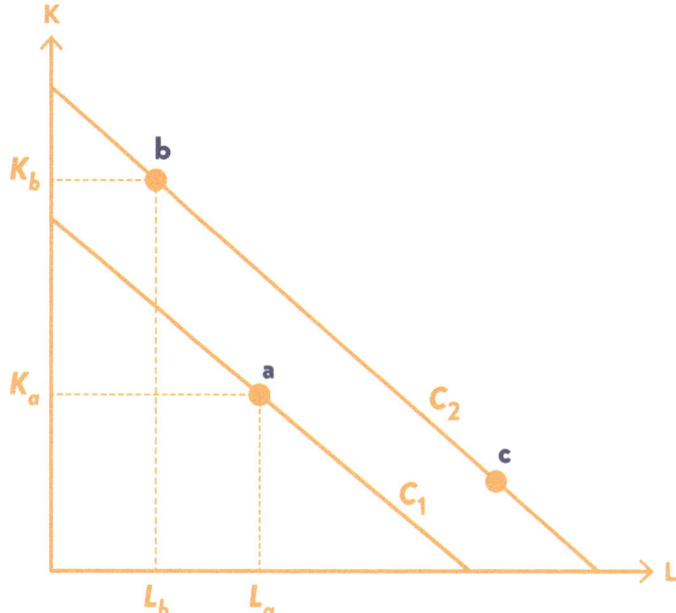

Fig. 5.2 Factor combinations and isocost lines. *Note* The two lines, C_1 and C_2 are *isocost lines*—combinations of the two inputs (labour and capital) that result in the same total cost. Costs along C_2 are higher than those along C_1, since it lies further out in the diagram. The slope of the isocost line is given by the ratio of factor prices, w/r. If the price of labour rises relative to the price of capital, the isocost line becomes steeper—and if the price of labour falls relative to capital, the isocost line becomes flatter

Labour costs consist of the wage per worker (w) multiplied by the number of workers (L), while capital costs are the price of capital (r) multiplied by the amount of capital (K). The wage level can be thought of as the annual salary for one full-time worker, while the price of capital is the annual cost of using one unit of capital (think of it as one machine) in the factory.

It doesn't matter whether you have paid for the capital equipment with your own money or borrowed someone else's money, for example from a bank. If you have borrowed the money, you must pay an annual interest. And if you have used your own money, it could alternatively have been placed in the bank and earned you interest income. Thus, capital has an opportunity cost (also called alternative cost or shadow price). The focus in this chapter is on labour and capital, so we ignore other costs such as raw materials and electricity but will return to those in the next chapter.

The isocost line is found by rearranging the costs and placing capital on the left side:

$$K = \frac{C}{r} - \frac{wL}{r} \quad \text{Isocost} \tag{5.2}$$

To find the slope of the isocost line, we take the derivative with respect to L:

$$-\frac{\partial K}{\partial L} = \frac{w}{r} \quad \text{Slope of isocost line} \tag{5.3}$$

Since the derivative expression is negative (to hire more workers, you must reduce the amount of capital to keep costs unchanged), the minus sign makes the left side positive. A change in the relative factor prices leads to a change in the slope of the isocost line. For example, an increase in the price of labour relative to the price of capital will make the isocost line steeper, while a decrease in the price of labour relative to capital will make the isocost line flatter.

5.4   Production Function and Isoquant

Conrad takes Anna's drawing back to Brian and explains that it does not seem profitable to choose the capital-intensive solution he had proposed, point b. But Brian is not giving up. He says there are many ways to produce one billion A4 sheets: one can choose various intermediate solutions between his suggestion and the current factor combination.

Brian explains that this can be thought of as a production function $Q(K, L)$ which shows how much output can be produced from all possible combinations of the input factors capital and labour.

$$Q = Q(K, L) \quad \text{Production function}$$

The production function in the paper mill takes the following form:

$$Q(K, L) = K^{0.5}L^{0.5} \text{ Balanced Cobb-Douglas production function} \tag{5.4}$$

This is an example of a Cobb-Douglas (CD) production function, where Q here represents the number of A4 sheets produced. The fact that the exponents are equal means that the two production factors are equally important in the production process: I call this a balanced Cobb-Douglas production function. We study a more general functional form in Math Box 5.1.

Brian throws off his jacket and rolls up his sleeves.

Brian: At the School of Engineering, we learned about the isoquant. It's a very useful tool when we think about choosing the combination of inputs. It shows all the combinations of production factors that yield a given level of output.

Conrad: Isoquant. Again, I get to enjoy my high school Greek! It means the same quantity. I'll remember that!

Labour and capital in the paper mill

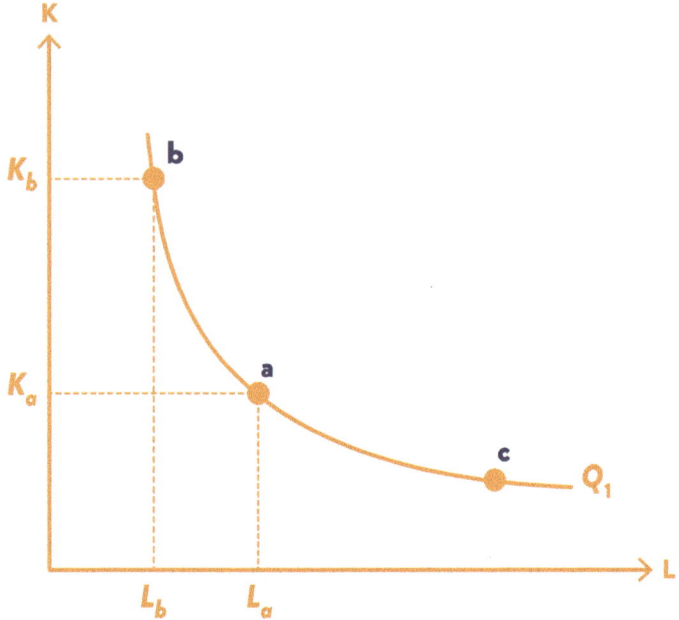

Fig. 5.3 Isoquant. *Note* The isoquant Q_1 shows different combinations of the two inputs that yield the same quantity produced. Points a, b and c, are possible combinations, but so are all other points on the isoquant. The slope of the isoquant is called the marginal rate of technical substitution (MRTS) and is given by the ratio of the marginal products of the two production factors

Figure 5.3 shows the isoquant that connects points a, b, and c. The slope of the isoquant is important—so important that it has its own name: the marginal rate of technical substitution, commonly abbreviated MRTS. It is important because it holds the key to understanding the firm's cost-minimising choice of production factors, which we will soon examine.

To understand the slope of the isoquant, we need to understand the concept of marginal product. This tells us how much output increases with a small increase in one of the production factors. Mathematically, the marginal product of capital, often abbreviated MP_K, can be expressed as the derivative of the production function with respect to capital input:

$$MP_K = \frac{\partial Q}{\partial K} \quad \text{Marginal product of capital} \tag{5.5}$$

Similarly, we can write the marginal product of labour as MP_L, which we find by differentiating the production function with respect to the input of labour:

$$MP_L = \frac{\partial Q}{\partial L} \quad \text{Marginal product of labour} \tag{5.6}$$

Why is the marginal product of the two factors important for understanding the slope of the isoquant? If the isoquant is steep, as in point b in Fig. 5.3, it means that one additional worker can replace many machines, which implies that the extra production contributed by the worker (i.e., the marginal product of labour) is much higher than the marginal product of capital. Conversely, at point c, where the slope of the isoquant is very flat, one additional worker cannot replace many machines without reducing production. This means that the marginal product of labour is lower than the marginal product of capital there.

Do you recognise the logic from the discussion about marginal utility and the slope of the indifference curve in consumer theory? See Sect. 1.3 for a refresher!

The marginal product of labour is higher at point b than at point c, and similarly, the marginal product of capital is higher at point c than at point b. This is intuitive: each worker becomes more productive when surrounded by more capital equipment, and each machine produces more when there is plenty of labour to operate it. A standard assumption is therefore that the marginal product of each production factor increases with the availability of the other factor, which we can express mathematically as:

$$\frac{\partial MP_K}{\partial L} > 0 \quad \text{Adding labour increases the marginal product of capital}$$

$$\frac{\partial MP_L}{\partial K} > 0 \quad \text{Adding capital increases the marginal product of labour}$$

Similarly, we assume that there are diminishing marginal products with respect to each factor. This means that as the input of one factor increases while the other remains fixed, the marginal product will decrease. This is also intuitive. Imagine a factory with one machine, and then we increase the labour input. After a while, it becomes crowded around the machine, and the production increase from adding one more worker becomes quite modest. Similarly, adding more machines in a firm with a fixed number of workers doesn't make much sense, since someone has to operate the new machines. We can express this mathematically as:

$$\frac{\partial MP_L}{\partial L} < 0 \quad \text{Decreasing marginal product of labour}$$

$$\frac{\partial MP_K}{\partial K} < 0 \quad \text{Decreasing marginal product of capital}$$

This is called the law of diminishing marginal product and plays a central role when we derive cost functions in the next chapter. So, remember this!

The slope of the isoquant can be derived using the implicit differentiation rule— the same one we used to find the slope of an indifference curve in the first chapter of the book. Along an isoquant, the quantity is constant; call it Q_1, and we start from the condition $Q(K, L) = Q_1$. By applying the implicit differentiation rule, we know that:

$$-\frac{dK}{dL} = \left(\frac{\partial Q}{\partial L}\right) / \left(\frac{\partial Q}{\partial K}\right)$$

The first expression on the right side of the equals sign is the marginal product of labour, MP_L, while the second one is the marginal product of capital, MP_K. The slope of the isoquant, the marginal rate of technical substitution (MRTS), can thus be written as follows:

$$\text{MRTS} = \frac{MP_L}{MP_K} \quad \text{The marginal rate of technical substitution} \qquad (5.7)$$

As with the corresponding expression in consumer theory, a useful rule of thumb is that the numerator in the MRTS gives the marginal product of the factor on the first axis (the horizontal one, measuring L), while the denominator gives the marginal product of the factor on the second axis (the vertical one, measuring K). The expression tells us how much capital can be given up if we increase labour input by one unit, without changing the level of production. We see that it is determined by the ratio of the marginal product of labour to the marginal product of capital.

In Math Box 5.1, we show in more detail how the MRTS relates to the marginal products of the two factors using a specific production function as an example.

Math Box 5.1 Isoquant
Assume the following production function:

$$Q(K, L) = K^\alpha L^{1-\alpha} \quad \text{General CD production function}$$

The exponent α indicates the relative importance of capital compared to labour in production: the higher α is, the more important capital becomes. We can find the marginal product of capital as:

$$MP_K = \frac{\partial Q}{\partial K} = \alpha K^{\alpha-1} L^{1-\alpha} = \alpha \left(\frac{L}{K}\right)^{1-\alpha} \quad \text{Marginal product of capital}$$

Similarly, the marginal product of labour is given by:

$$MP_L = \frac{\partial Q}{\partial L} = (1 - \alpha) K^\alpha L^{-\alpha} = (1 - \alpha) \left(\frac{K}{L}\right)^\alpha \quad \text{Marginal product of labour}$$

We notice that MP_K increases with L and decreases with K: the marginal product of capital rises as labour input increases, and falls as capital input increases, in line with the discussion above. Similarly, MP_L increases with

K and decreases with L. If we divide the marginal product of labour by the marginal product of capital, we get:

$$\frac{MP_L}{MP_K} = \frac{(1-\alpha)\left(\frac{K}{L}\right)^{\alpha}}{\alpha\left(\frac{L}{K}\right)^{1-\alpha}} = \frac{(1-\alpha)}{\alpha}\frac{K^{\alpha}L^{-\alpha}}{L^{1-\alpha}K^{\alpha-1}} = \frac{(1-\alpha)}{\alpha}\frac{K}{L}$$

The marginal rate of technical substitution can thus be expressed as:

$$MRTS = \frac{MP_L}{MP_K} = \frac{(1-\alpha)}{\alpha}\frac{K}{L} \quad \text{Marginal rate of technical substitution}$$

With a balanced CD production function ($\alpha = 0.5$) this simplifies to:

$$MRTS = \frac{MP_L}{MP_K} = \frac{K}{L}$$

5.5   Cost Minimisation

Conrad feels a bit wiser and asks Anna whether it might be a good idea to change production slightly, pointing especially to point d in Fig. 5.4, which represents a compromise between Brian's recommendation at point b and the current point a.

But Anna says no again. She explains that point a provides the absolute best balance between capital and labour on the isoquant Q_1. This is the isoquant here is tangent to the isocost line. Or, in other words: the slopes of the isocost and isoquant are equal. All other factor combinations along Q_1 would lead to higher costs.

Since the slope of the isoquant is given by the marginal rate of technical substitution (MRTS), and the slope of the isocost is given by the factor price ratio, the condition for a cost-minimising factor combination can be expressed as:

$$MRTS = \frac{MP_L}{MP_K} = \frac{w}{r} \quad \text{Optimal input combination} \qquad (5.8)$$

We have here used the graphical approach to find the optimum. From Fig. 5.4 we know that the optimum is defined by the tangency point between the isocost line and the isoquant, which means their slopes are equal. We then derive the slopes of the two and set them equal, and, as we have seen, this implies that the ratio of marginal products must equal the factor price ratio.

Notice how similar this is to consumer theory. Consumers' goal is to get as much utility as possible for their money. We saw, for example, how a change in prices rotates the budget line and how the indifference curve moves in response to that, and we analysed substitution and income effects.

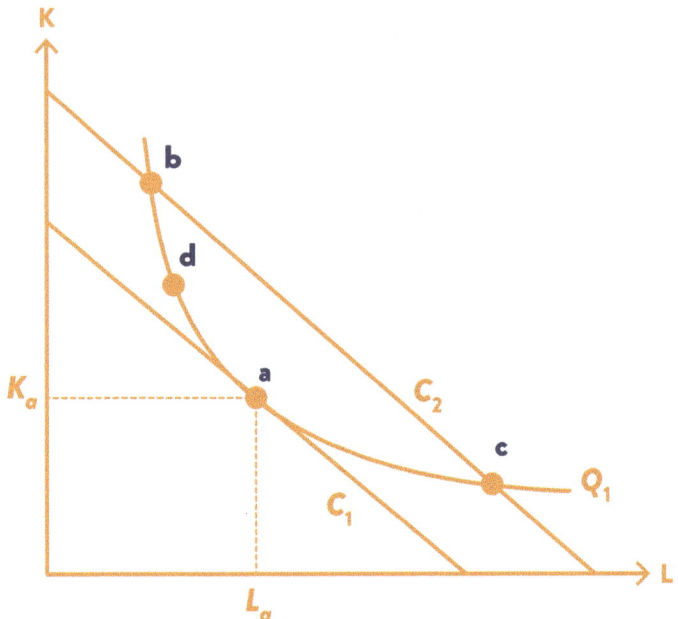

Fig. 5.4 Cost minimisation. *Note* The figure shows an isoquant Q_1 and two isocost lines, with $C_1 < C_2$. Point a, with capital K_a and labour L_a is the cost-minimising combination for isoquant Q_1. Here, the isoquant and isocost are tangent—in other words, the slope of the isoquant equals the slope of the isocost. Any other point along the isoquant would incur higher costs

Producers' goal is slightly different: they must ask, "How can we produce a given output at the lowest possible cost?" Here the quantity is fixed, and costs adjust to that quantity, with the optimum occurring where the isocost is tangent to the isoquant. So, whereas utility maximisation in consumer theory means that the curve (the indifference curve) moves to the line (the budget line), cost minimisation in producer theory means that the line (the isocost) moves to the curve (the isoquant)! This also means that in producer theory there is no discussion of substitution and income effects, and we do not use concepts like "normal inputs" or "inferior inputs."

As in consumer theory, you can also find the optimal factor combination using the Lagrange method, so let's briefly go through it here. We can find the cost-minimising combination of the production factors K and L, for a given quantity $\overline{Q}$, by formulating the Lagrange-expression:

$$\mathcal{L} = wL + rK - \lambda(Q(K, L) - \overline{Q})$$

The first two terms are the costs, while the expression in parentheses is the production function minus a fixed output level $\overline{Q}$, and λ is the Lagrange multiplier. The Lagrange method here is about minimising costs for a given output. Relating the math to the figure, this means starting from a given isoquant and finding the isocost line that is tangent to it. We begin by differentiating the Lagrangian with respect to L and then with respect to K, and setting those derivatives equal to zero:

$$\frac{\partial \mathcal{L}}{\partial L} = w - \lambda \frac{\partial Q}{\partial L} = 0$$

$$\frac{\partial \mathcal{L}}{\partial K} = r - \lambda \frac{\partial Q}{\partial K} = 0$$

Note that in the second term on the right-hand side, the Lagrange multiplier is multiplied by the marginal product: for labour, $(\partial Q/\partial L) = MP_L$, and for capital, $(\partial Q/\partial K) = MP_K$. By combining the first two first-order conditions, we therefore recover the optimal factor-combination condition in (Eq. 5.8): $MRTS = MP_L/MP_K = w/r$. We then differentiate the Lagrangian with respect to λ and set this derivative equal to zero:

$$\frac{\partial \mathcal{L}}{\partial \lambda} = Q(K, L) - \overline{Q} = 0$$

This gives us the condition $Q(K, L) = \overline{Q}$, i.e., that we remain on the same isoquant. It is useful to demonstrate the optimal choice using a concrete production function. We do this in Math Box 5.2. But before that, here is a recipe that shows the path to the optimum:

▶   **Recipe Number 3. Three Steps to the Optimal Factor Combination**

Step 1. Compute the MRTS
Use the specific production function to derive the marginal rate of technical substitution (MRTS), defined as the ratio of the marginal products of the two inputs:
Step 2. Apply the Tangency Condition
Set your expression for MRTS equal to the ratio of input prices, $MRTS = w/r$. Solve this equation to find the optimal capital–labour ratio, K/L.
Step 3. Solve for Each Input
Solve the optimal factor ratio for one of the factors, and substitute this expression into the production function. From this you can find the optimal use of that one factor, which can then be plugged into the optimal factor ratio to find the optimal input level of the other factor as well.

Math Box 5.2 Cost Minimisation

Assume the general Cobb-Douglas production function:

$$Q(K, L) = K^{\alpha} L^{1-\alpha} \quad \text{General CD production function}$$

The exponent α indicates the relative importance of capital compared to labour in production:

The higher α is, the more important capital is.

Step 1 in the recipe above is about finding the MRTS based on the specific production function.

Note the similarity in method to the one we used to find optimal consumption in Chapter 1—see Recipe 1. From Math Box 5.1, we know that with the given production function we have:

$$MRTS = \frac{MP_L}{MP_K} = \frac{(1-\alpha)}{\alpha} \frac{K}{L} \quad \text{Marginal rate of technical substitution}$$

And that completes step 1. In step 2, we use this expression together with the condition for the optimal combination of inputs MRTS = w/r, and find:

$$MRTS = \frac{(1-\alpha)}{\alpha} \frac{K}{L} = \frac{w}{r}$$

A simple rearrangement of this expression gives us the optimal input ratio:

$$\frac{K}{L} = \frac{w}{r} \frac{\alpha}{(1-\alpha)} \quad \text{Optimal input ratio}$$

This tells us that production should be more capital-intensive the higher is the price of labour (w) relative to the price of capital (r), and the more important capital is in the production process (the higher α is). If $\alpha = 0.5$, i.e., a balanced Cobb-Douglas production function, the optimal input ratio simplifies to:

$$\frac{K}{L} = \frac{w}{r}$$

Or put differently:

$$rK = wL$$

We see that with a balanced Cobb-Douglas production function, cost minimisation means spending the same amount of money on each input factor.

That concludes step 2. Step 3 says that we should solve the optimal input combination for one of the inputs, and we solve for capital:

$$K = \frac{w}{r}\frac{\alpha L}{(1-\alpha)}$$

We insert this in the production function and find:

$$Q = K^\alpha L^{1-\alpha} = \left(\frac{w}{r}\frac{\alpha L}{(1-\alpha)}\right)^\alpha L^{1-\alpha}$$

This can be expressed as:

$$Q = \left(\frac{w}{r}\frac{\alpha}{(1-\alpha)}\right)^\alpha L$$

With a bit of manipulation, we find the optimal input of labour as:

$$L = Q\left(\frac{(1-\alpha)}{\alpha}\frac{r}{w}\right)^\alpha \quad \text{Optimal input of labour}$$

To find the chosen level of capital, we insert the optimal amount of labour into the optimal input ratio expressed in terms of capital input:

$$K = \frac{w}{r}\frac{\alpha}{(1-\alpha)}Q\left(\frac{(1-\alpha)}{\alpha}\frac{r}{w}\right)^\alpha$$

This can be rearranged as follows:

$$K = Q\left(\frac{\alpha}{(1-\alpha)}\frac{w}{r}\right)\left(\frac{\alpha}{(1-\alpha)}\frac{w}{r}\right)^{-\alpha}$$

Which means that:

$$K = Q\left(\frac{\alpha}{(1-\alpha)}\frac{w}{r}\right)^{1-\alpha} \quad \text{Optimal input of capital}$$

Notice that with a balanced CD production function ($\alpha = 0.5$) the optimal input of labour and capital simplifies to:

$$L = Q\sqrt{\frac{r}{w}}$$

$$K = Q\sqrt{\frac{w}{r}}$$

We can insert these expressions of optimal factor use in the cost function:

$$C = wL + rK$$

With a general CD production function the total costs are then:

$$TC = w\left(\frac{(1-\alpha)}{\alpha}\frac{r}{w}\right)^{\alpha}Q + r\left(\frac{\alpha}{(1-\alpha)}\frac{w}{r}\right)^{1-\alpha}Q$$

And with a balanced CD production function ($\alpha = 0.5$) this simplifies to:

$$TC = w\left(\frac{r}{w}\right)^{0.5}Q + r\left(\frac{w}{r}\right)^{0.5}Q = \left(2\sqrt{wr}\right)Q$$

5.6   Changes in Relative Factor Prices

Conrad has just returned from the annual paper conference in Stuttgart, where he heard that many companies in the paper industry are considering relocating to China. Labour is cheap there—he's heard that wages are only a quarter of what they are at home. So surely the calculation is simple, Conrad thinks, but just to be sure, he checks with Anna.

Conrad: I've been thinking about China, Anna. What if we move production there?

Anna: Still a billion sheets?

Conrad: Yes, a billion. That way, we could cut labour costs to a quarter, right?

Anna: Hold on, not so fast. We can do even better than that! With the lower wages in China, we shouldn't produce in the same way as we do here. Let me show you with a diagram.

Anna picks up a pencil and a piece of paper. "We can divide the cost savings into two parts. First, there are savings to be made by doing exactly what we do now, but in China, where labour is cheaper. That's what you had in mind with your calculation, Grandpa. But second, there's a gain from choosing a more labour-intensive production technique, like I showed you in the diagram, where we increase the use of labour and reduce the use of capital compared to how we do it in the paper mill here at home."

Figure 5.5 shows the optimal choice in the *home* country at point a where the isocost line C_a, based on domestic factor prices, is tangent to the isoquant.

If Conrad moves production to China without changing the factor intensity of production—that is, still operating at point a on the isoquant—this results in costs C_b', where the flatter isocost line in China reflects the lower wage level there. If the price of capital is the same in China as at home, this means unchanged capital costs but lower labour costs, indeed, only a quarter of the labour costs at home, if Conrad is right.

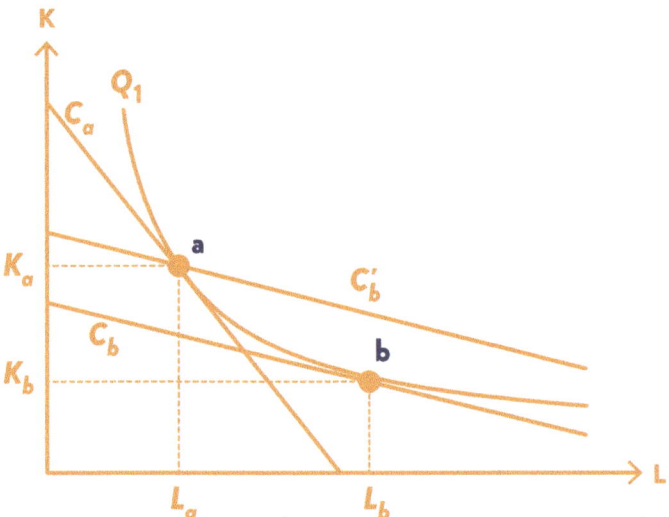

Fig. 5.5 Optimal input combination with different wage levels. *Note* The figure shows the optimal input combination for a given isoquant when the price of labour (w) is relatively high, as represented by the isocost line C_a, and when it is low, as represented by the isocost line C_b. We see that the tangency point between the isocost and the isoquant differs in the two situations. With high w, the firm chooses point a, while with low w, the firm opts for the more labour-intensive production at point b. If the firm had moved production to China but kept the same input mix as at home, this would have resulted in costs C_b'

But Anna points out that there are even greater savings to be made by relocating to China. Point a is not a cost-minimising combination of inputs in China; that is found at point b, where the isocost line C_b, which has the same slope as C_b', is tangent to the isoquant. This involves an increase in labour and a reduction in capital along the isoquant, that is, a shift to a more labour-intensive production method compared to the one used at home.

The cost saving from optimising input use in line with local factor prices is given by the difference between C_b and C_b', and that's the opportunity Conrad had overlooked.

"Yes, there's more to be saved by moving to China than I thought," says Conrad.

"But you know, money isn't everything! I'm fond of this factory and the people who work here."

Conrad decides to take a nap and think things over.

5.7   Four Production Technologies

While Conrad takes his afternoon nap, Brian reflects on the microeconomics lectures from his first year at the School of Economics. He didn't always pay close attention, but he found technology fascinating.

Brian remembers that the professor once told a story about going to the hairdresser (he remembered this example particularly well, since the professor had so little hair that Brian wondered what he was doing there at all), and how he asked the students to consider the production technology in a hair salon, where, unsurprisingly, haircuts are produced using a hairdresser and a pair of scissors.

"What does the isoquant look like in a hair salon?" the professor asked. For the first time in the course, Brian raised his hand.

Brian: Normally, we think there's some degree of flexibility in the mix of production factors, but that's not the case at the hairdresser's.

Professor: That's right, go on!

Brian: You need one hairdresser and one pair of scissors to cut hair. A pair of scissors without a hairdresser is no help, and you won't get more haircuts by giving the same hairdresser an extra pair of scissors—unless the hairdresser's name is Edward, that is.

Professor: Edward?

Brian: Yes, Edward Scissorhands. Have you seen the film?

Professor: No, but I see what you mean. Funny. But for ordinary hairdressers, what do you think the isoquant looks like?

Brian: I think it's L-shaped.

Professor: Very good. Let me show that more formally on the board.

The situation in the hair salon can be expressed as follows:

$$Q = \min(K, L) \quad \text{Perfect complementarity between } K \text{ and } L$$

This is called a Leontief production function, named after the Russian-American economist and Nobel laureate Wassily Leontief (1905–1999). In this case, the input that is in shortest supply always determines the amount produced. In other words, there is no flexibility in the combination of production factors. For example, you can't compensate for a lack of capital by using more labour—each worker needs a certain amount of equipment to be productive.

Panel A of Fig. 5.6 illustrates the isoquant for this production function. The firm must use at least L_a workers and K_a units of capital to produce this output, and cost minimisation means choosing exactly this combination of inputs, point a, regardless of relative factor prices.

The second story Brian remembers from the same lecture was when the professor told them about a trip to the shop. He ended up somewhere with only self-checkout tills, something he had never seen before, and it worked brilliantly. Once again, the professor asked the students to consider the isoquant to produce "*checkout services in a shop.*"

Brian remembered that Anna had raised her hand this time, partly because he knew she worked part time in a grocery shop, and partly because he thought she was very pretty. Anna had said that she believed the isoquant would be linear, and she was absolutely right, as Panel B of Fig. 5.6 illustrates. A shop can choose to use only staffed checkouts (point b), only self-checkouts (point a), or a mix of

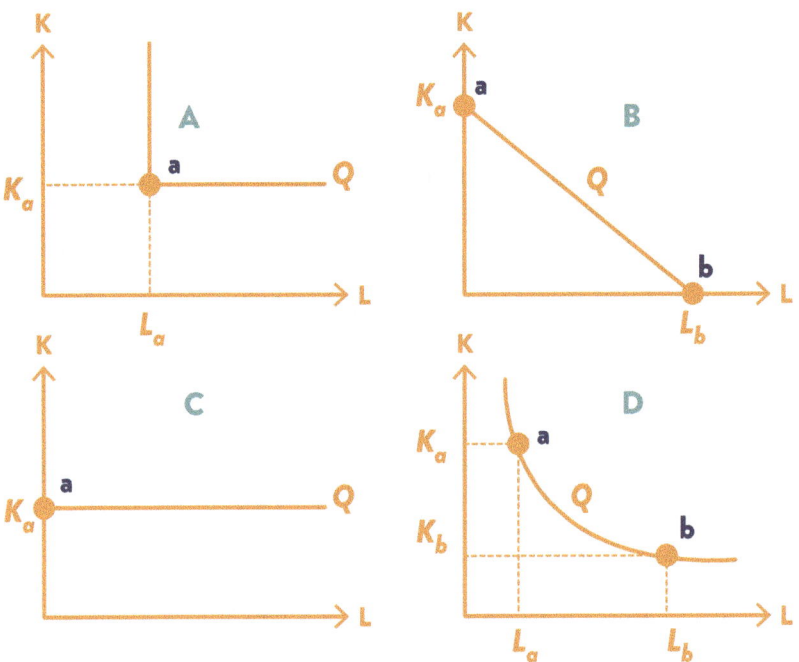

Fig. 5.6 Isoquants with different technologies. *Note* Panel A shows the case of Leontief technology, where the inputs are perfect complements: the isoquant is L-shaped. Panel B illustrates the case of linear technology, where the inputs are perfect substitutes: the isoquant is a straight line. Panel C shows robot technology, where production takes place exclusively using capital: K_a indicates the amount of robot capital required, and the more efficient the robots are, the lower is K_a. Panel D displays the isoquant based on a Cobb-Douglas production function

both (a point somewhere between *a* and *b* on the isoquant). With perfect substitutability between inputs, the producer will always choose to use only the cheapest input. And if both inputs cost the same, then the combination doesn't matter.

The third production technology the professor discussed involved robots. Brian remembered this example well, because he was more interested in machines than in people (apart from Anna, of course). The professor explained that one could think of robots as a kind of *super capital*, able to operate entirely on their own, see Panel C of Fig. 5.6. In this case, the isoquant is a horizontal line: production is automated, and labour is no longer a relevant input.

How many robots are needed to produce a given quantity depends on how efficient they are—and of course, the newest robots are the best: they are super-duper capital! Automation is one of the major issues of our time. More on that shortly.

For comparison, the professor had also included the isoquant from a balanced Cobb-Douglas production function, as illustrated in Panel D. As we can see, this represents something in between the Leontief technology in Panel A and the linear technology in Panel B.

To summarise: With Cobb-Douglas technology, the firm always uses both inputs but has flexibility in how to combine them (e.g., it may choose a combination at point *a* or *b*). With Leontief technology, the firm also always uses both inputs, and ideally in fixed proportions. With linear technology, the firm uses only one input, the cheaper one. And finally, with robot technology, there is only one relevant input: capital.

5.8   Robots or Traditional Technology?

"Robots," thinks Brian, "that's the future, A4 paper production should be automated, the sooner the better."

Brian bursts into Conrad's office, bringing with him a strange robotic creation he's put together in his spare time.

"Boss, this is important, it's about a whole new way of doing things. We need to think AI instead of A4. I mean, we need to think outside the box!"

Conrad, who has just finished his afternoon nap, stares in astonishment at Brian's robot.

"You see, Conrad, robots can do the whole job! Well, not the one I built, but it gave me the idea. It's about artificial intelligence, computers, and automation, and it's happening now."

"It's complicated, but it can be simplified. Think of robots as super capital," Brian explains. "The machines can do the job entirely on their own!"

$$Q = AK \qquad \text{Production function with robot technology}$$

The factor A in the production function represents the efficiency of robot capital. The higher A is, the more A4 sheets each robot produces.

Brian's robot

We can rewrite the production function to express the capital requirement in production. Suppose we want to produce a quantity Q_1, then the number of robots needed to produce this amount can be written as:

$$K = \frac{Q_1}{A} \equiv R_1$$

We see that the more efficient the robots are, that is, the higher A is, the fewer machines are needed to produce a given quantity. In panel C of Fig. 5.6, a higher A would imply shifting the Q-line downwards.

Brian draws Fig. 5.7 to illustrate the choice between traditional and new technology. The traditional way of producing one billion sheets is represented by the isoquant Q_1^T. Points a and b represent two different ways of producing a billion A4 sheets using traditional technology. Point R_1 shows how much capital is needed to produce the same quantity using robots. More efficient robots will move point R_1 down along the vertical axis: fewer robots are needed to do the same job.

"You know, Conrad, I have long advocated for more capital-intensive production. But we should think completely differently! Why choose b when we can choose robots?"

Conrad listens with interest. "They talked quite a lot about robotisation at the paper conference in Stuttgart, so I understand this is a current trend, but let's hear what Anna has to say—she's good at spotting the economics in projects like these."

Anna studies Brian's drawing and begins diplomatically: "I completely agree that robots are better than point b with traditional technology: the same capital requirement but lower labour costs, that calculation is straightforward."

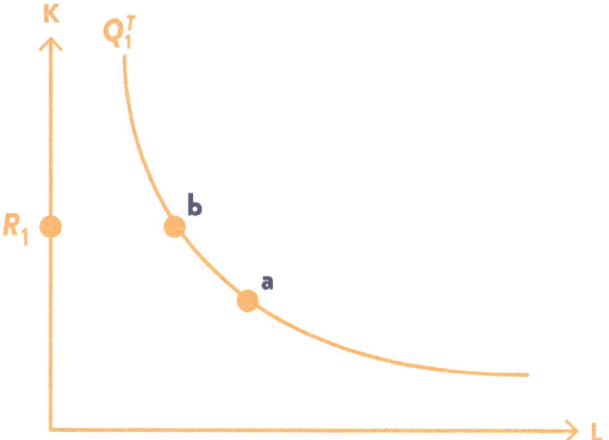

Fig. 5.7 Robot and traditional technology. *Note* The figure shows the isoquant Q_1^T for traditional technology and the capital requirement to produce the same quantity with robots, given by R_1. Points a and b show different input combinations to produce the same quantity using traditional technology

"But" she continues, "it's not obvious that robotisation is a better alternative than the current production at point a. Automation certainly means saved labour costs, but it requires investments, and capital isn't free."

Anna draws the isocost line C_a based on today's factor prices, as shown in Fig. 5.8. Initially, the paper mill has made the optimal choice of point a, where the isoquant is tangent to the isocost line. She calls the point where the isocost hits the vertical axis K_1^a, and adds: "With today's factor prices, and current robot technology, there's actually no gain from automating production."

How can she see this? The cost-minimising input combination is given by point a, with capital input K_a and labour input L_a. If we move along the isocost line C_a up to the point where it crosses the vertical axis, we get a measure of production costs (with traditional technology) expressed in units of capital (since $L = 0$ along the vertical axis). We call this point K_1^a.

At the same time, we know that R_1 is the capital requirement with robot production. If $K_1^a = R_1$, then the cost of producing Q_1 must be the same for the two technologies.

Of course, costs are not always equal for the two technologies. In Fig. 5.9, we conduct two thought experiments. It's challenging to show two things in one figure, but let's try!

Start with the isocost line C_a, and then assume that labour becomes more expensive. As we saw from the discussion above, Conrad was initially indifferent between traditional technology and robot technology. So, in one way, the answer is obvious: labour has become more expensive, and Conrad should now choose to automate with robots.

But let's look at this a bit more formally. With the increase in the wage level w, the isocost line becomes steeper, C_b, and the best way to produce one billion

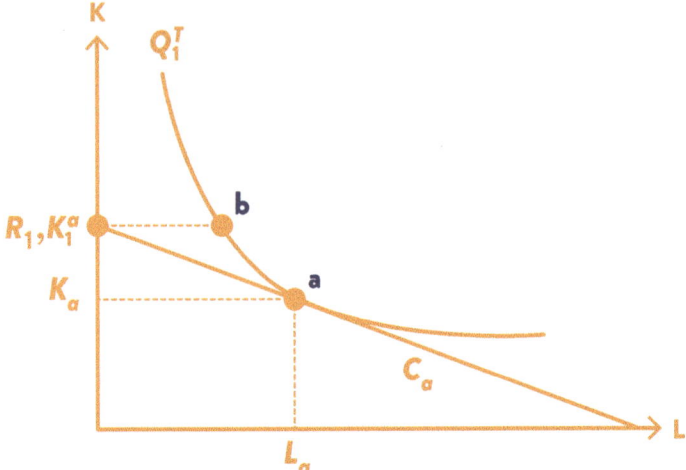

Fig. 5.8 Robot or not? *Note* The figure shows the isoquant Q_1^T for traditional technology and the capital requirement R_1 needed to produce the same quantity using robot technology. The optimal combination with traditional technology is at point a, and C_a is the isocost line for this combination. The isocost line crosses the vertical axis at point K_1^a, which represents costs measured in units of capital. In this case, $K_1^a = R_1$, meaning the cost is the same for both traditional and robot technology.

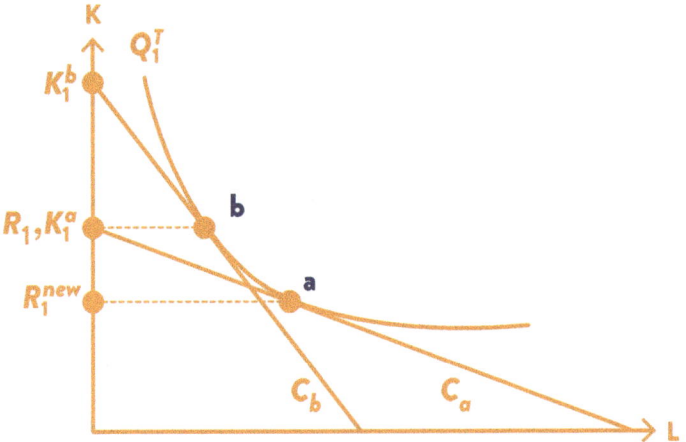

Fig. 5.9 Higher wages, more efficient robots, and technology choice. *Note* The figure shows how an increase in wages (w) shifts the isocost line from C_a to C_b. The optimal combination using traditional technology changes from point a to point b. The higher wage makes it profitable to switch to robot technology since $K_1^b > R_1$. Similarly, more efficient robots reduce the capital requirement in production from R_1 to R_1^{new}. This also makes choosing robots profitable, as $K_1^a > R_1^{new}$

A4 sheets using traditional technology is now at point b (a more capital-intensive production, as you can see). The cost of production using traditional technology, measured in units of capital, is now K_1^b. Since $K_1^b > R_1$, we can conclude that producing with robots is cheaper.

The second thought experiment concerns more efficient robots. Let's return to the starting point with the original wage level and isocost line C_a, where Conrad is indifferent between traditional technology and automation using the good old robots.

A new generation of robots means an increase in A, and the capital required per unit produced decreases, marked in the figure as R_1^{new}. Again, it's obvious what Conrad should choose: if he was initially indifferent, then more efficient robots must mean he chooses automation, since $K_1^a > R_1^{new}$.

And if you study the figure more closely, the choice of technology is obvious. The capital requirement with the more efficient robots is the same as with traditional technology (in the figure given by point a), so capital costs remain equal, while robotisation saves labour costs. The answer is clear: Conrad should automate.

5.9 Summary

We have studied how the inputs labour and capital can be combined to produce a good, in Conrad's factory, A4 paper. The possible combinations can be expressed mathematically as a production function.

A key concept here is the isoquant, which shows all the combinations of labour and capital that produce the same output quantity. The slope of the isoquant is called the marginal rate of technical substitution (MRTS) and is given by the ratio of the marginal products of the two inputs.

The cost side of production is summarised by the isocost line, which shows all the combinations of the two inputs that incur the same cost. The producer minimises costs by choosing the combination where the isocost line is tangent to the isoquant.

We have also seen how changes in factor prices affect this choice—for example, lower labour costs lead to a more labour-intensive production.

We have studied different technologies: some require labour and capital in fixed proportions, while others allow more flexibility in the input mix than the Cobb-Douglas production function implies.

Automation of production, where the entire process is carried out by machines, is an exciting and topical issue. Brian is a technology enthusiast eager for robotisation in the paper mill, but Anna shows that the choice between robots and traditional technology depends on factor price ratios and how efficient the robots are: you should not automatically automate!

5.10   Key Terms

Isocost: Shows combinations of two inputs that yield the same total cost.

Production function: Shows how much output can be produced from different combinations of inputs.

Marginal product: The increase in output when the input of one factor is increased by one unit.

Isoquant: Shows combinations of inputs that produce the same quantity of output.

Law of diminishing returns: The assumption that the marginal product of an input decreases as the quantity of that input increases.

Marginal rate of technical substitution (MRTS): Shows the slope of the isoquant, that is, how much less capital is needed if the input of labour is increased by one unit, without reducing output.

Perfect complements in production: When production requires inputs of labour and capital in a fixed proportion. Also called a Leontief production function.

Perfect substitutes in production: When production can be carried out equally well using only labour or only capital.

5.11   Multiple-Choice Exercises

5.1: Isocost
Assume a firm has daily production costs of 8,000 euros, with the wage rate (w) at 200 euros per day and the price of capital (r) at 40 euros per day. The firm's isocost line is then given by:

A. C1
B. C2
C. C3
D. C4

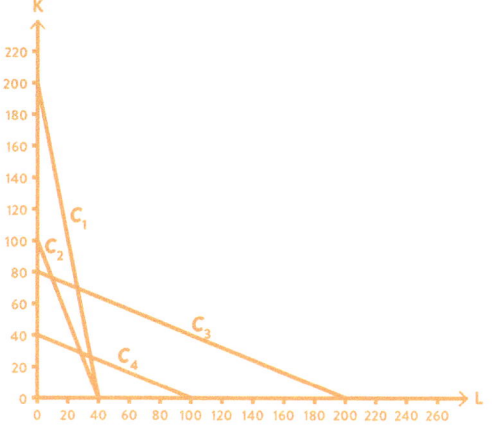

5.2: Isoquant

Consider Fig. 5.3, reproduced here. Which of the following statements is/are correct?

1. More is produced at point b than at point c.
2. It is cheaper to produce at point b than at point c.
3. The marginal rate of technical substitution (MRTS) between labour and capital is the same at points b and c.

A. 1, 2 and 3
B. Only 1
C. Only 3
D. Neither 1, 2 or 3

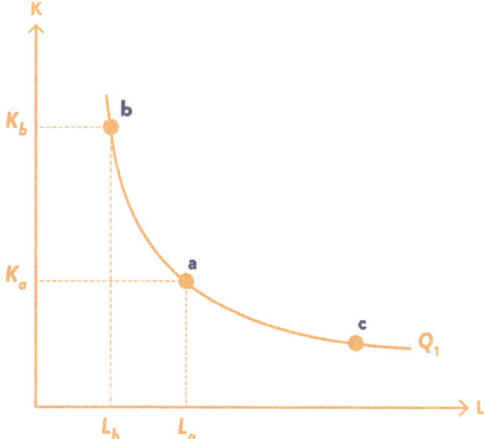

5.3: Cost Minimisation

Assume the firm's initial combination is at point *a* the figure below. Suppose there is a change in factor prices so that the cost-minimising combination on the isoquant Q_a is now at point *b*. What could have happened to the factor prices?

A. The wage *w* has increased, no change in *r*
B. The price of capital *r* has decreased, no change in *w*
C. The wage *w* has increased, and the price of capital *r* has decreased
D. All of the above

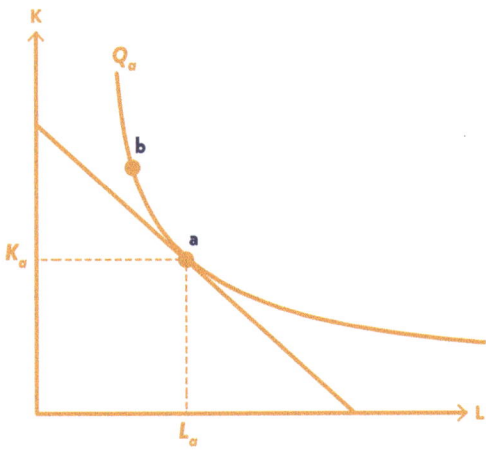

5.4: Other Production Technologies

Assume a firm with perfect substitutability between capital (*K*) and labour (*L*). The isoquant is given in the figure below. The price of capital (*r*) is 100 euros, while the price of labour (*w*) is 90 euros. At which point on the isoquant will the firm choose to minimise its costs?

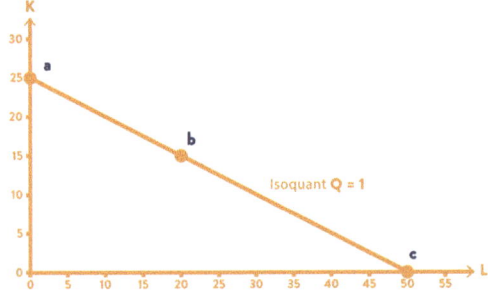

A. a
B. b
C. c
D. It doesn't matter: the costs are the same regardless

Solutions: 5.1 A; 5.2 D; 5.3 D; 5.4 A.

Costs

6

Conrad wants to increase production and wonders what it would take. Brian explains that it depends on the time horizon.

Anna introduces Conrad to marginal cost, and says that it is the most important cost of all!

6.1 Introduction

Conrad realises that production is cheaper in China and that robots might be the future, but he has ambitions for the old paper mill, which he inherited from his father. He is considering increasing production—in fact, doubling it!

Conrad dreams of expansion

© The Author(s), under exclusive license to Springer Nature Switzerland AG 2026   105
K. Bjorvatn, *Microeconomics Made Simple*, Classroom Companion: Economics,
https://doi.org/10.1007/978-3-032-06354-0_6

He has a colleague with a larger and more modern production facility, more employees, and much higher output, and Conrad thinks about what his colleague has achieved and whether he should do something similar himself.

But what does it take to expand, and what will it cost? Fired up, he grabs his walking stick, gets up from his armchair, and strides briskly to Brian's office.

Conrad: Brian, I want to produce more! Twice as much!

Brian: Haha, I assume you've been talking to your colleague again! When were you thinking?

Conrad: As soon as possible!

Brian: Everything's possible, but you know the factory has certain limitations. It'll take at least two years to get new machines in place, and we may also need larger production facilities.

Conrad: I understand, but I'm an old man and don't have time to wait. How much can we increase production right now?

Brian: Then we'd have to hire more people, and that's no problem—though it might get a bit crowded.

Conrad: OK, but what will it cost?

From Conrad and Brian's dialogue, we see that the time horizon is a key factor when considering expansion. In the paper mill, it takes two years to get new equipment in place, though this will naturally vary from one industry to another. It takes less time to acquire new capital equipment in a taxi company (new cars) than in a paper mill. Regardless, the time horizon is a central consideration in most industries.

In production theory, we therefore distinguish between the short run and the long run. In the short run, capital is fixed, which means the only way to expand is by hiring more workers. In the long run, however, new capital equipment can be brought in, which opens new possibilities.

We begin with the short-run perspective. How many workers are needed to increase production when the capital stock is fixed, and what does it cost? We'll mainly use standard production functions, but also consider situations where there is perfect complementarity between inputs, implying that producers can meet an absolute capacity constraint. We'll also address the challenge of optimising production across different production facilities, where marginal thinking becomes crucial.

The chapter concludes with the long-run perspective, where all production factors are variable. The discussion of costs, and how the time horizon affects them, is most relevant when we focus on traditional technology and the interaction between labour and capital. So, in this chapter, we'll set the robots aside.

6.2   Labour Requirements and Costs in the Short Run

Conrad asks Brian for an estimate of how much labour is needed to double production. Brian illustrates this with Fig. 6.1. We are currently at point a, with a production level of Q_1, one billion sheets of A4 paper, using L_a labour and K_a capital. Q_2 shows a doubling of production, and to achieve this with the factory's current capital stock, we need L_c workers.

"As you can see, Conrad, this requires a quite large increase in the number of employees. In fact, an exponential increase." Brian demonstrates this based on the paper mill's production function, $Q = K_0^{0.5} L^{0.5}$, where K_0 is the installed capital. To make the point even clearer, he simplifies by setting $K_0 = 1$, which makes the production function $Q = L^{0.5}$. From this, we can easily find the labour requirement as $L = Q^2$. Doubling Q from 1 (billion sheets) to 2 (billion sheets) requires quadrupling the labour input!

"The reason for this," says Brian, "is the law of diminishing returns" (see the previous chapter, I told you it would be important!). "Each new worker we hire produces less than the one before. And since each worker receives the same wage, expanding production gets rather expensive."

Conrad realises that doubling production in the short run may be overly ambitious, but he goes to Anna to enquire what it would cost to increase output—even if

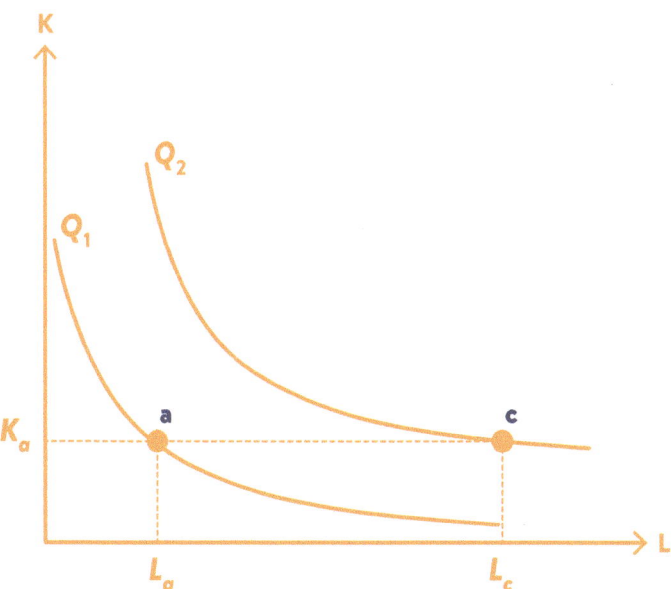

Fig. 6.1 Labour Requirement for Production Increase. *Note* The figure takes as point of departure a production level Q_1 with factor combination given by point *a*. For a given capital input, an increase in labour from L_a to L_c is required to produce Q_2.

not twice as much, then at least a bit. He shows her Brian's figure. Anna examines it and adds the isocost lines, based on today's factor prices, as shown in Fig. 6.2.

We currently have an optimal input combination at point *a*. You can see this because the isocost line is tangent to the isoquant Q_1 at that point. The cost here is given by C_1. Doubling output moves us to point *c*, and thus to a much higher cost, represented by the isocost line C_3.

This input combination does not represent the long-run cost minimum—that is found at point *b*, where the input of capital and labour is more balanced. But achieving that requires increasing capital from K_a to K_b, and that will take time.

"Let's look at what it costs to expand a little or a lot, so we have a better basis for decision-making," says Anna. There are three important cost measures we need to keep an eye on:

1. Marginal cost (*MC*)
2. Average variable cost (*AVC*)
3. Average total cost (*ATC*)

Of these, marginal cost is by far the most important! Why? Because it's the key to understanding how to achieve maximum profitability in production. But the other

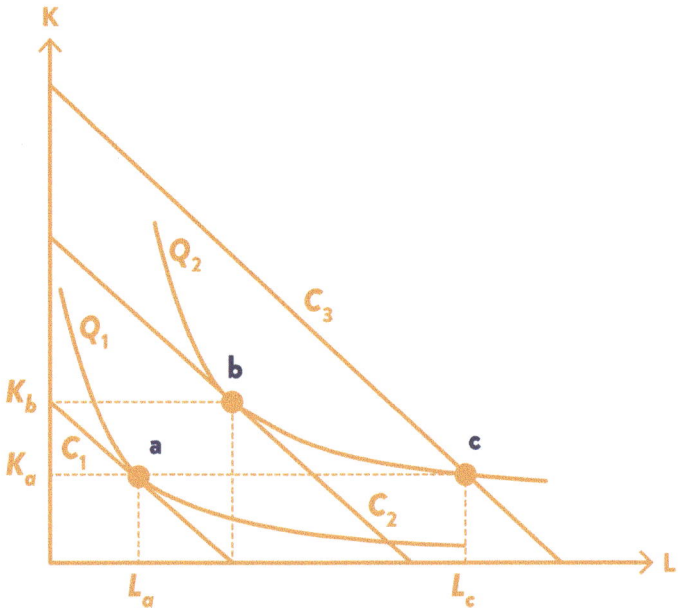

Fig. 6.2 Input combination and production increase, short and long run. *Note* The figure starts with production Q_1 and choice of point *a*. A short-run increase to Q_2 requires raising labour input to L_c, with costs given by C_3. In the long run, one would choose the combination at point *b*, with costs C_2, which are lower than C_3.

two are also useful for assessing whether production is in fact profitable. That will be the topic of the next chapter.

To move forward, we need to one step back, to the definition of costs. In the previous chapter, we focused on the costs associated with labour and capital (see Eq. 5.1), but now we also include the costs of raw materials (broadly defined to include electricity and any other inputs required in production), which we denote by Z:

$$C = wL + rK + Z \quad \text{Costs} \tag{6.1}$$

The first term on the right-hand side is labour costs, the second is capital costs, and the third is raw material costs. Since capital is fixed in the short run (K_0), we can split costs into variable and fixed. The variable costs (VC) are those that vary with output, that is, they are linked to labour and raw materials:

$$VC = wL + Z \quad \text{Variable costs} \tag{6.2}$$

Fixed costs are linked to capital:

$$FC = rK_0 \quad \text{Fixed costs} \tag{6.3}$$

An important point when discussing costs in economic theory versus how they're treated in business accounting is the distinction between economic costs and accounting costs. The difference can be summed up in one word: *opportunity cost* (also known as *alternative cost*). In a company's accounts, only actual expenditures are recorded, whereas in economic theory it isn't necessary to spend any money at all for something to count as a cost. For example, economic theory treats capital as having a cost even if the firm owns all of the machines and has no debt, since that capital has an alternative use—the machines could be sold or rented out.

But what if there are no buyers for the capital equipment, perhaps because it's custom-built for this firm? The capital costs are then irreversible, what we call a sunk cost. In such cases, the capital has no alternative value, so its opportunity cost is zero.

Opportunity cost doesn't apply only to capital. If you're self-employed, your labour time has a cost even if you don't pay yourself a wage: you could alternatively have taken paid employment elsewhere.

We are now ready to look at the three key cost measures that Anna said we must watch, starting with the most important: marginal cost (MC). This tells us how much total cost increases when output increases by one unit, and is defined as:

$$MC = \frac{\partial C}{\partial Q} \quad \text{Marginal cost} \tag{6.4}$$

The second most important cost is the average variable cost (AVC):

$$AVC = \frac{VC}{Q} = \frac{wL + Z}{Q} \quad \text{Average variable cost} \tag{6.5}$$

And finally, the average total cost (ATC):

$$ATC = \frac{C}{Q} = \frac{wL + Z + rK_0}{Q} \quad \text{Average total cost} \tag{6.6}$$

In Math Box 6.1, we study these three key cost functions using a specific production function.

Math Box 6.1 Short-Run Cost Functions
Assume the production function:

$$Q = K_0^{0.5} L^{0.5}$$

This can be rearranged to give the labour requirement in production as:

$$L = \frac{Q^2}{K_0}$$

Labour costs can therefore be expressed as a function of output:

$$wL = \frac{wQ^2}{K_0} \quad \text{Labour costs}$$

Assume that each unit produced requires one unit of raw material (broadly defined to include electricity), and that the price of raw material is z. The raw-material cost is then:

$$Z = zQ \quad \text{Cost of raw materials}$$

The total costs can therefore be written as:

$$C = wL + Z + rK_0 = \frac{wQ^2}{K_0} + zQ + rK_0$$

Here, the first two terms on the right side are the variable costs:

$$VC = wL + Z = \frac{wQ^2}{K_0} + zQ \quad \text{Variable costs}$$

The fixed costs are:

$$FC = rK_0 \quad \text{Fixed costs}$$

The marginal cost can be found by differentiating the costs with respect to Q (it does not matter whether you start from total or variable costs—think about why!):

$$MC = \frac{\partial C}{\partial Q} = \frac{2wQ}{K_0} + z \quad \text{Marginal cost}$$

We can find the average variable cost by dividing the variable costs by the produced quantity:

$$AVC = \frac{VC}{Q} = \frac{wQ}{K_0} + z \quad \text{Average variable cost}$$

To find the average total costs, we first define the average fixed cost (AFC):

$$AFC = \frac{rK_0}{Q} \quad \text{Average fixed cost}$$

The average total cost is therefore:

$$ATC = AVC + AFC = \frac{wQ}{K_0} + z + \frac{rK_0}{Q} \quad \text{Average total cost}$$

There are three important points to note about the cost functions derived in Math Box 6.1.

First, both the marginal cost (MC) and the average variable costs (AVC) increase with the quantity produced, and MC rises faster than AVC. This is because when capital is fixed, production can only be increased by increasing labour input. Since the marginal product of labour is diminishing while the wage rate w remains constant, the cost of producing the last unit is higher than the previous one. This means that both marginal cost and average variable costs rise, but marginal cost increases more than the average costs.

Second, since the quantity Q appears in both the numerator and denominator in the average total cost (ATC), the ATC curve takes a U-shape. At low production levels, fixed costs dominate, so increasing production lowers the average fixed costs (AFC), and thus the average total costs fall. At higher production levels, variable costs dominate, and the rising AVC pulls the total average cost curve upwards.

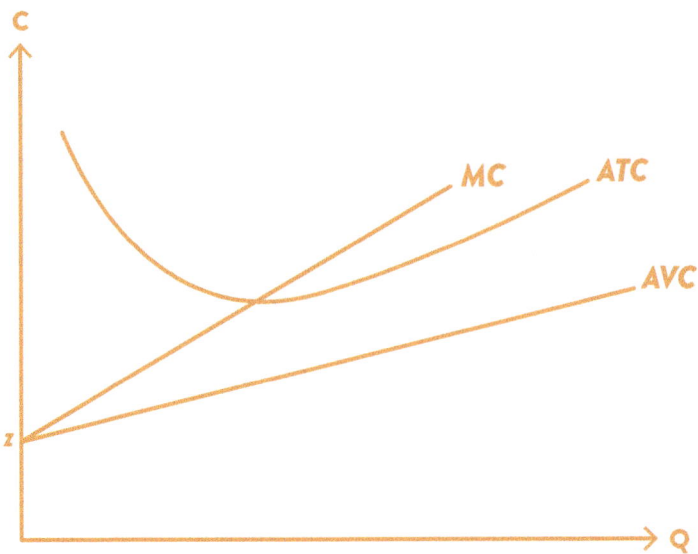

Fig. 6.3 Short-run cost functions. *Note* The figure shows the marginal cost (*MC*), average variable cost (*AVC*), and average total cost (*ATC*) as functions of the quantity produced *Q*. The *MC* and *AVC* curves are rising because labour becomes less productive as the quantity produced increases, given a fixed amount of capital. The *ATC* curve is U-shaped because fixed costs dominate at low production levels, and average fixed costs decline as production increases, causing *ATC* to fall. For higher production levels, variable costs dominate, and the rising *AVC* pulls the *ATC* curve upward

Third, the price of raw materials *z* determines the level of the cost functions: For any given quantity produced, a higher *z* shifts the cost curves upward, while a lower *z* shifts them downward.

Figure 6.3 illustrates the short-run cost functions in the paper mill, based on what we found in Math Box 6.1.

6.3   When the Producer Hits the Wall

So far, we have used Cobb-Douglas production technology to derive cost functions. This is a technology where capital and labour work together, and firms have significant flexibility in choosing their combination of inputs. This provides a realistic description of production in many firms. However, as we discussed in Sect. 5. 7, some production processes have more, and some less, flexibility in combining capital and labour than what is built into the Cobb-Douglas function.

It is especially interesting to consider the case of a Leontief production function. We remember that this is a technology that uses production factors in fixed

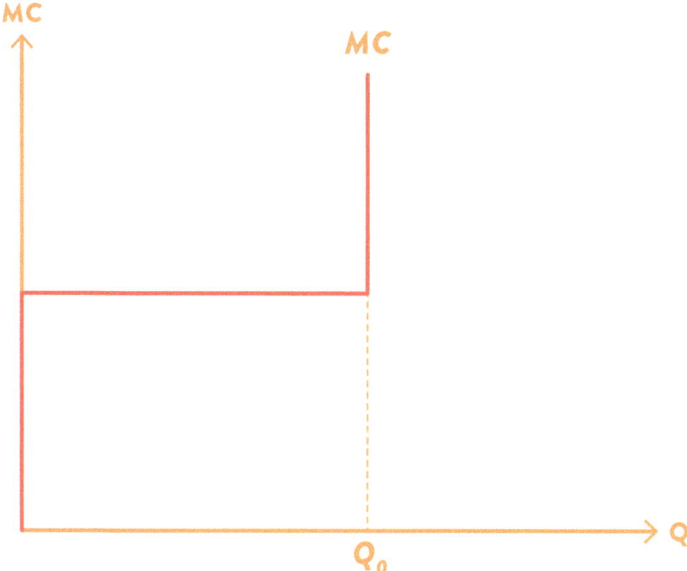

Fig. 6.4 Marginal Costs with Perfect Complementarity. *Note* The figure shows the marginal cost with a production function where the input factors are perfect complements, meaning they must be used in a fixed ratio. For a low production level, in the figure for $Q < Q_0$, capital is not a limiting factor, and the marginal cost up to this point is constant. At Q_0, the installed capital becomes a binding constraint: the marginal cost therefore becomes vertical at this point

proportions, like a hairdresser and a pair of scissors. But it can also be a hydro-electric power plant where water is the variable input, and the size of the dam defines a maximum output. Or a gas power plant driven by gas and turbines.

What do the cost functions look like when the factors must be combined in fixed proportions? The answer is: like a staircase. See Fig. 6.4. The height of each step is given by the marginal cost, and the step stops where production reaches the capacity limit, Q_0, determined by the installed capital. The capital defines a production ceiling, which causes the marginal cost to skyrocket. The producer hits a wall, in the sense that it is physically impossible to produce a larger quantity. The staircase in this figure consists of only one step but imagine a power producer who has both hydroelectric and gas power. Then the staircase will consist of several steps. Water is free, and this is the first step, while gas is expensive, and this is the second step. Later in the book (see Sect. 8.7), we will take a closer look at the electricity market.

6.4   Cost Minimisation with Multiple Production Units

Conrad has two paper factories; one located in the south of the county and the other further north. The two plants have different technologies: Production in the south can best be described by a Cobb-Douglas function, with costs as described in Fig. 6.3, while the northern one has a Leontief function, with costs as shown in Fig. 6.4. Conrad has decided to produce a given amount of paper, one billion sheets per year, but is uncertain about how to distribute production between the two plants. He has therefore called Anna and Brian to the office for consultation.

The flipchart with the robot analysis from the day before is still there, and Anna turns the sheet and begins to draw (Fig. 6.5). "Imagine a bathtub," she says. "We measure production from the southern factory from left to right and production from the northern factory from right to left; these are the ends of the bathtub, while the bottom represents the total production level, which is fixed."

Production in the south is given by Q_S with marginal costs MC_S and average variable costs AVC_S. Production in the north is Q_N with marginal costs MC_N and equally high average variable costs AVC_N: this is the horizontal part of the cost function. For simplicity, we assume that the capacity limit is not binding (there is more than enough installed capacity at the northern plant to produce one billion sheets per year). Brian argues that the production should be allocated to give equal average cost in the two plants. This means a configuration at point b, where production in the south is Q_b and in the north is $Q - Q_b$. Conrad, on the other hand, thinks it must be most reasonable to distribute production evenly between the two plants, at point c: Both factories exist, and there is no reason not to use them equally.

Brian argues that average costs should be equalized (he's wrong)

But Anna knows better. "We have to think marginally!" she says, referring to point *a* where the marginal cost in the southern factory crosses that in the northern factory. Production in the south becomes Q_a, and in the north $Q - Q_a$. She explains that this allocation will yield cost savings corresponding to area C compared to Conrad's proposal, and then additional savings B compared to Brian's proposal. Notice that point *b* has a much higher marginal cost in the south than in the north: the last unit could therefore have been produced a lot more cheaply in the north, which makes it sensible to shift some production there—meaning a movement to the left in the bathtub. Each unit moved from the south to the north yields a cost saving equal to the difference between the two marginal costs, and in total the savings equal the area B + C when moving from Brian's proposal to Anna's. Marginal thinking can save big money! This is one of the deepest insights in producer theory.

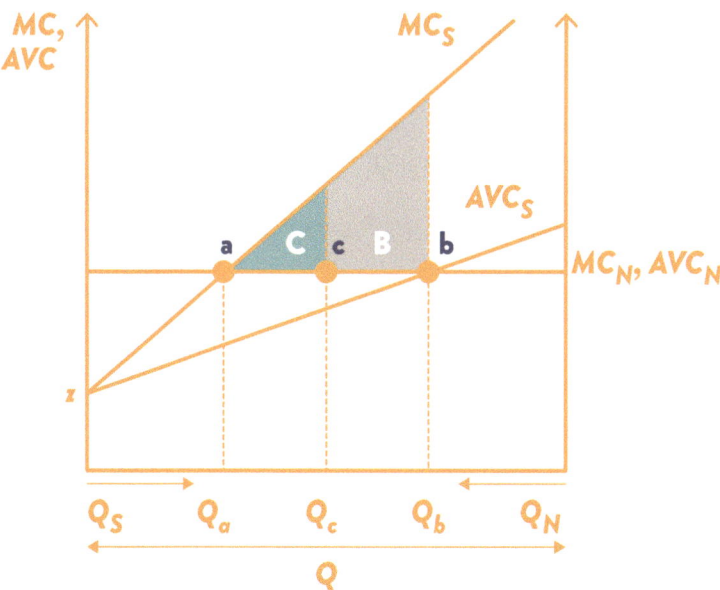

Fig. 6.5 Cost minimisation with multiple production units. *Note* Production in the southern plant is given by Q_S, with marginal costs MC_S and average variable costs AVC_S, where $MC_S > AVC_S$. Production in the northern plant is Q_N, and for this plant, marginal and average variable costs are the same: $MC_N = AVC_N$. Total output is given by Q. Equal production in both plants corresponds to point *c*. Producing at equal average costs leads to point *b*, while equalising marginal costs gives point *a*. Moving from point *b* to *c* results in a cost saving represented by area *B*, while moving from *c* to *a* yields an additional saving equal to area *C*

6.5 Factor Requirements and Costs in the Long Run

In the short run, higher production typically leads to higher marginal costs, as implied by the law of diminishing returns. In the long run, however, this law no longer applies. The factor requirement for expanding production depends on the so-called returns to scale—what happens to output when both labour and capital inputs are scaled up.

This is intuitive. If both labour and capital are doubled, total costs will double too. The effect on average costs, however, depends on how this scaling affects output. If output also doubles, this is referred to as constant returns to scale, and the average cost remains unchanged. If output increases by less than double, we are in a situation of decreasing returns to scale, which means average costs rise. But if output more than doubles, we have increasing returns to scale, and average costs fall. Falling average costs are also referred to as economies of scale.

Math Box 6.2 provides a more detailed mathematical description of the returns to scale in production, based on specific production functions.

Math Box 6.2 Economies of Scale

Here we will present three production functions that lead to three different outcomes in terms of returns to scale, starting with Cobb-Douglas.

$$Q_1 = K^{0.5}L^{0.5}$$

What happens to production if we double the input of capital and labour? Let's call the new production level Q_1', and we find:

$$Q_1' = (2K)^{0.5}(2L)^{0.5} = 2K^{0.5}L^{0.5} = 2Q_1$$

As we can see, doubling both factors also doubles production. This is an example of a production function with constant returns to scale. All Cobb-Douglas production functions where the exponents sum to one have this property.

But what about the following production function?

$$Q_2 = KL$$

What happens to production if we double the input of capital and labour here? Let's call the new production level Q_2', and we find

$$Q_2' = (2K)(2L) = 4KL = 4Q_2$$

Here we see that doubling the input factors has led to a quadrupling of production, which is an example of a production function with increasing returns to scale. This generally applies to Cobb-Douglas functions where the sum of the exponents is greater than one.

Finally, let's look at the following production function:

$$Q_3 = K^{0.25}L^{0.25}$$

Again, we double the input of capital and labour and call the new production level Q_3':

$$Q_3' = (2K)^{0.25}(2L)^{0.25} = 2^{0.5}K^{0.25}L^{0.25} = \sqrt{2}Q_3$$

In this case, we see that doubling the input factors results in a production increase that is less than double, an example of decreasing returns to scale. As you might have guessed, this applies to all Cobb-Douglas functions where the sum of the exponents is less than one.

From Math Box 6.2, we see that the production function we have used consistently throughout this part of the book, where the exponents sum to one, has constant returns to scale. We know that, based on this, the long-run average costs will be constant. In Math Box 6.3, we present this more formally.

Math Box 6.3 Long-Run Cost Functions

Let's start with the general CD production function:

$$Q = K^{\alpha}L^{1-\alpha}$$

We found in Math Box 5.2 in the previous chapter that the optimal input of capital and labour in this case is given by:

$$K = Q\left(\frac{\alpha}{(1-\alpha)}\frac{w}{r}\right)^{1-\alpha} \quad \text{Optimal input of capital}$$

$$L = Q\left(\frac{(1-\alpha)}{\alpha}\frac{r}{w}\right)^{\alpha} \quad \text{Optimal input of labour}$$

We insert these expressions into the cost function:

$$C = wL + rK + Z$$

And find that the long-run costs, TC_{LR}, can be written as:

$$TC_{LR} = w\left(\frac{(1-\alpha)}{\alpha}\frac{r}{w}\right)^{\alpha}Q + r\left(\frac{\alpha}{(1-\alpha)}\frac{w}{r}\right)^{1-\alpha}Q + zQ \text{ Long} - \text{run cost function}$$

This is the firm's long-run cost function with a general Cobb-Douglas production function. Let us simplify and assume a balanced Cobb-Douglas production function ($\alpha = 0.5$), which gives us:

$$TC_{LR} = w\left(\frac{r}{w}\right)^{0.5}Q + r\left(\frac{w}{r}\right)^{0.5}Q + zQ = \left(2\sqrt{wr} + z\right)Q$$

The long-run average costs, ATC_{LR}, are then given by:

$$ATC_{LR} = \frac{TC_{LR}}{Q} = 2\sqrt{wr} + z$$

We see that the average costs are not affected by the quantity produced: they remain constant. Similarly, the long-run marginal cost, MC_{LR}, is given by:

$$MC_{LR} = \frac{\partial TC_{LR}}{\partial Q} = 2\sqrt{wr} + z$$

We observe that in this case, $ATC_{LR} = MC_{LR}$ and independent of Q. This is because this technology exhibits constant returns to scale, as discussed in Math Box 6.2.

6.6   Comparison of Short-Run and Long-Run Cost Functions

We have now derived the cost functions for both the short run and the long run, giving us a good understanding of what it costs to increase production. However, we have not yet considered the short-run and long-run costs together, so let's do that now. Figure 6.6 illustrates this.

There are several interesting points to note about this figure. First, we see that the short-run marginal cost (MC) is lower than the long-run marginal cost (MC_{LR}) when production is low. Why is this the case, you make ask. Isn't it easier to keep costs low in the long run than in the short run? To understand this result, we must remember that the short-run marginal cost does not take capital costs into account. These are not relevant in an analysis of what it costs to increase production in the short run, since capital is fixed in that time frame. The low short-run marginal cost at low production volume can be explained by there being few workers and a lot of capital. One could say that the capital stock is oversized relative to the

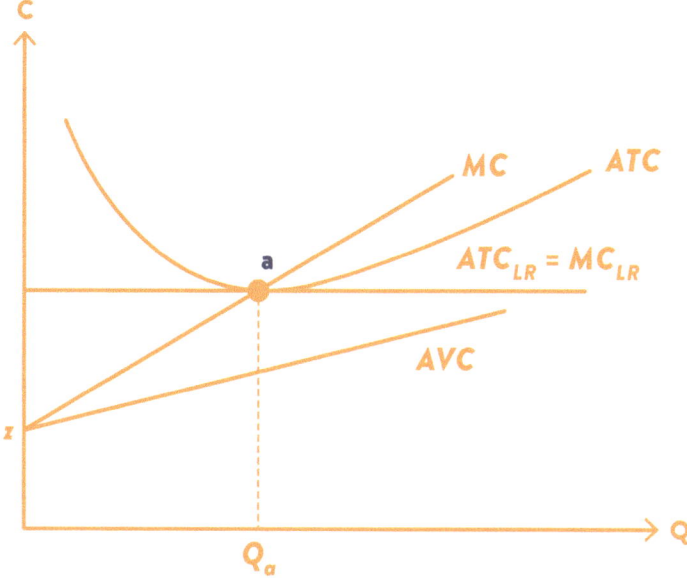

Fig. 6.6 Cost functions, short and long run. *Note* The figure shows the cost functions for the short run and the long run. We see that the short-run marginal cost (*MC*) equals the long-run marginal cost (MC_{LR}) at point *a*, corresponding to production level Q_a. At this quantity, the installed production capital in the short run, K_0, is also optimal in the long run. For production levels less than Q_a, the installed capital is oversized, which makes labour very productive and thus the short-run marginal cost low. For production levels greater than Q_a, the installed capital is undersized, causing the short-run marginal cost to be high. The short-run total average cost (*ATC*) curve touches the long-run cost curve at the production level Q_a, where the installed capital in the short run is also optimal for the long run

current production. This makes labour very productive, resulting in low short-run marginal costs.

Now that we understand this, we also understand why the short-run marginal cost lies above the long-run marginal cost when production is high, shown in the figure as above Q_a. When production volumes are high, in the short run we have many workers working with relatively little capital (you can think of it as congestion around the paper machines). The installed capital in this case is undersized, making labour less productive and thus increasing the marginal cost.

We also observe that the two marginal cost curves intersect at point *a*, corresponding to production level Q_a. Based on what we've just explained, you might understand why this is the case: here, the installed capital is exactly what would also be optimal in the long run. In other words, the short-run capital stock is perfectly sized for that production level.

Long-run costs can be seen as a benchmark for short-run costs: a large difference between them suggests there may be potential to reduce costs by adjusting the level of capital, either scaling up or down. The goal must be a combination of labour and capital that minimises costs, as described in the previous chapter.

However, as we have seen in this chapter, this is a process that can take time. In the short run, one may have to live with a factor combination that deviates from the long-run optimum, resulting in short-run costs that are higher than long-run costs.

6.7   Summary

In this chapter, we have examined what happens to resource needs and costs when a producer changes production levels. We have studied three cost concepts: marginal cost, average variable cost, and average total cost. These costs are central when we in the next chapter address profitability: marginal cost because it contains information about which production level the firm should choose to maximise profit, and the two average costs to assess whether the firm is in fact profitable.

Of these three costs, marginal cost is the most important. This was clear when we analysed a firm's production allocation between two facilities: choose the allocation that equalizes marginal costs. When we study the profit maximising level of production in the next chapter, we will again see that marginal cost is central.

The focus of this chapter has been on the short-run perspective, where installed capital is fixed: this is the reality firms face in their ongoing production decisions. But it is also interesting to consider the long-run perspective, where capital can be adjusted.

The different time perspectives give rise to different cost patterns. The flexibility firms have in the long run allows total costs to be kept lower than in the short run. In the short run, marginal cost will necessarily increase with the quantity produced, due to the law of diminishing marginal returns: each new worker contributes less to production than the previous one when capital is fixed. This applies at least with standard technology such as Cobb-Douglas, but as we have seen, Leontief technology presents a somewhat different picture, with constant marginal cost up to a capacity limit. In the long run, however, there is no law of diminishing marginal returns, and the cost function depends on the scale properties of production. A Cobb-Douglas production function where the exponents sum to one exhibits constant returns to scale, and in this case the long-run costs—both average and marginal—are constant.

6.8   Key Terms

Short run: Capital is fixed.
Long run: All production factors are variable.
Variable cost: Costs that vary with the quantity produced.
Fixed cost: Costs that do not vary with the quantity produced.
Opportunity cost or alternative cost. The value of time or money in its best alternative use
Sunk cost: A cost that cannot be recovered or changed, even in the long run.

Marginal cost: The increase in costs when production rises by one unit.
Constant returns to scale: When doubling the input factors also doubles the output.
Decreasing returns to scale: When doubling the input factors results in less than double the output.
Increasing returns to scale: When doubling the input factors results in more than double the output.

6.9   Multiple-Choice Exercises

6.1: Short-Run Cost Function

In Fig. 6.3, we see that in the short run, the average total costs (ATC) first fall before eventually rising. Why do they fall?

A. Because there are more units over which to spread the energy costs (Z)
B. Because there are more units over which to spread the fixed costs
C. Because the marginal product of labour is diminishing
D. Because of all three reasons mentioned above

6.2: Costs with Other Production Technologies

Hairdressers and scissors are an example of production with Leontief technology, where each hairdresser needs one pair of scissors. Imagine an exclusive hair salon with a fixed number of specially made scissors from Japan (it takes a long time to order new ones): Why would the marginal cost then be constant up to a certain production level?

A. Because the salon initially has too many hairdressers
B. Because the salon does not minimise costs in the short run
C. Because production is only limited by the number of hairdressers
D. Because there are constant returns to scale in the short run

6.3: Returns to Scale

What are the returns to scale of the production function $Q = K^{0.5}L$?

A. Constant returns to scale
B. Increasing returns to scale
C. Decreasing returns to scale
D. First increasing, then decreasing returns to scale

6.4: Long-Run Cost Function

What happens to long-run average costs if the production function has increasing returns to scale?

A. They will be U-shaped
B. They will fall
C. They will rise
D. They will be constant

Solutions: 6.1 B; 6.2 C; 6.3 B; 6.4 B.

Profit

7

Anna explains that it can be profitable to increase productioneven if the company is not especially profitable...

... and that it can make sense to reduce productionnow but expand in the longer term. Conrad asks for an explanation.

7.1 Introduction

As we've seen, Conrad wants to increase the production of paper. In the previous chapter, we explored what this would cost. But ultimately, the key question is: is it profitable?

Conrad heads to the bank to discuss the profitability of scaling up paper production

© The Author(s), under exclusive license to Springer Nature Switzerland AG 2026   123
K. Bjorvatn, *Microeconomics Made Simple*, Classroom Companion: Economics,
https://doi.org/10.1007/978-3-032-06354-0_7

To answer this, we must first define profitability. There are two central concepts: operating profit and profit. In the short run, the relevant measure is the operating profit. In the long run, the firm must also generate a positive profit—or at the very least, avoid a negative one.

We'll see that the firm is at its most profitable when production is at a level where marginal cost equals price (remember how I said marginal cost is the most important one?). But being most profitable doesn't necessarily mean the firm is in fact profitable! We'll look at the thresholds a firm needs to meet to survive, both in the short run and in the long run.

Finally, we'll examine the firm's demand for labour. Production requires workers, and changes in prices and wages affect both how much a profit-maximising firm should produce, and how many people it should employ. By looking at the relationship between a firm's demand for labour and supply of goods, we get a more complete picture of how the firm operates.

7.2 Profitability Defined: Profits and Operating Profits

Conrad wants to increase production, and the meeting with Anna and Brian has given him a better understanding of what it takes (in terms of labour and capital) and what it costs. Now he's wondering whether increasing production is profitable.

To help answer this, Anna starts by defining profitability. She explains that the factory is profitable if it can pay for labour and raw materials and still provide the owners of capital with a return on their investment that they're satisfied with.

Raw materials have a price, and workers' wages are also set in the market. But what will make capital owners satisfied?

Here we again need to distinguish between the short run and the long run. In the short run, capital is fixed, so capital owners don't have much room to act. What matters for the firm in the short run is whether it can generate enough income to pay the workers and cover the cost of raw materials. In other words, the income must cover the variable costs (VC), and the relevant profitability measure is operating profits.

Operating profits (OP) can be expressed as:

$$OP = PQ - VC \quad \text{Operating profits} \qquad (7.1)$$

Here, revenue is given by price P multiplied by quantity Q, and VC refers to the variable costs. Without a positive operating profit, the firm is unprofitable even in the short run and should shut down immediately. A positive operating profit means that the firm not only covers its variable costs, but that its income also contributes to covering at least some of the fixed costs.

But imagine that the operating profit is zero or just barely positive. Yes, the income is enough to cover the variable costs, but there's not much left for the capital owners. They're unlikely to be pleased. And this is not a situation that can last.

If the operating profit is poor, the owners might sell the machines and other capital equipment and instead place their money in the bank (which we can think of as the best available alternative). Conrad's paper mill is a family business, but there is a minority of outside investors, and they might start making noise if profitability is too low. And even though Conrad is fond of the factory, he too would prefer to see his investment yield a return.

Conrad pays his workers

Long-run profitability is about keeping the capital owners satisfied. And for them to be satisfied, the money they have invested in the factory must yield a return that is at least as good as what they could have earned by putting the money in the bank at an interest rate r. I simplify by setting aside the issue of risk associated with different investments: think of the factory as being as safe as the bank.

The opportunity cost of tying up K_0 in the factory is therefore rK_0. As we saw in the previous chapter, we referred to this as the firm's fixed costs.

$$FC = rK_0 \quad \text{Fixed costs}$$

In the long run, the revenue generated by the factory must therefore cover both the variable and the fixed costs, which together make up the total costs. We say that the firm must generate a profit. Mathematically, profit (π) can be expressed as follows:

$$\pi = PQ - VC - FC \quad \text{Profit} \tag{7.2}$$

Profits are therefore operating profits minus fixed costs:

$$\pi = OP - FC \quad \text{Profit and operating profit} \tag{7.3}$$

If the operating profits are greater than the fixed costs, then production in the factory yields a return higher than what the owners could have earned by putting their money in the bank. We say the factory generates an economic profit, $\pi > 0$, and the owners are naturally very satisfied.

If the operating profits are exactly equal to the fixed costs, then the return in the factory is the same as what they could have earned by placing the money in the bank, and the owners are reasonably satisfied. This is zero economic profit, $\pi = 0$, which simply means the profitability is normal.

If the operating profits are lower than the fixed costs, however, the owners are dissatisfied. Now profit is negative, $\pi < 0$, and if this situation persists, they will sell their shares in the company. In the long run, the company will therefore close down.

Remember that by costs we mean *opportunity costs*, as discussed in the previous chapter. For example, if capital costs are irreversible (sunk costs), then the opportunity cost is zero. In this case, operating profits remain the relevant profitability measure even in the long run: The owners must simply accept whatever they get! If operating profits are poor and profit is negative, the owners may wish they had never invested in the business—but it's too late to change that!

7.3   Profit Maximisation

Now that Anna has explained the different profitability concepts to Conrad, she is ready to address the question of whether the paper mill should increase production, and if so, by how much. She begins with the assumption that the firm aims to maximise its profit. The production level that maximises profit is found by taking the derivative of profit (or of operating profits—the result is the same) with respect to quantity and setting it equal to zero:

$$\frac{\partial \pi}{\partial Q} = P - \frac{\partial VC}{\partial Q} = 0 \Rightarrow P = MC \quad \text{Profit maximisation condition} \qquad (7.4)$$

As we can see, profitability is maximised when production is at a level where marginal cost equals price. Intuitively, profit is maximised when the revenue from the last unit equals its cost. When the last unit produced costs less than what it can be sold for, it is worthwhile to produce it, whereas if the last unit costs more than the price, production should be reduced to increase profitability. This is a simple but very important observation, as it guides the decisions of profit-maximising firms in a perfectly competitive market.

However, maximum profitability does not necessarily mean the firm is profitable, so it is also important to note the break-even points for profitability:

$$\pi = 0 \Rightarrow PQ - VC = FC \Rightarrow OP = FC \quad \text{Zero profit} \qquad (7.5)$$

That is, when the operating profit is just enough to cover the fixed costs, the profit is zero. Remember that this means the capital owners receive a normal return on their investment. Similarly, the operating profit is zero when the following condition is met:

$$OP = 0 \Rightarrow PQ = VC \quad \text{Zero operating profit} \tag{7.6}$$

Here, the firm's revenue is just sufficient to cover the variable costs, that is, the labour expenses and raw material costs. In Math Box 7.1, we show a firm's profit-maximising choice of quantity based on a specific production function and discuss profitability.

Math Box 7.1 Profit maximisation and profitability

Starting from a balanced Cobb-Douglas production function $Q = K_0^{0.5} L^{0.5}$, we know from Math Box 6.1 that the marginal cost is given by:

$$MC = \frac{\partial C}{\partial Q} = \frac{2wQ}{K_0} + z \quad \text{Marginal costs}$$

A firm maximises profit by choosing a quantity such that $P = MC$, which with the marginal cost above means:

$$P = \frac{2wQ}{K_0} + z$$

With simple manipulation, we can find the optimal quantity produced as:

$$Q = \frac{(P - z)K_0}{2w} \quad \text{Optimal quantity}$$

While this is the profit-maximising choice of production, we don't know whether production is in fact profitable. So, let's investigate that, and for simplicity do so based on a numerical example where $w = r = z = K_0 = 1$. From Math Box 6.1 profit can in this case be expressed as:

$$\pi = OP - FC = PQ - Q^2 - Q - 1 \quad \text{Profit}$$

Using the quadratic formula for second-degree equations $ax^2 + bx + c = 0$, where in the profit expression above $a = -1, b = P - 1$, and $c = -1$, we find that profit is zero for

$$Q = \frac{P - 1 \pm \sqrt{(P - 1)^2 - 4}}{2}$$

Profit can thus be zero at both low and high production levels. Zero profit at low production is due to fixed costs being spread over very few units, while zero profit at high production is due to high variable costs. Note that for $P = 3$, profit is zero only at one production level, namely $Q = 1$, while for $P < 3$, profit is never positive at any level of Q.

Figure 7.1 summarises the discussion so far. It builds on what we learned about cost functions in the previous chapter, but here we add two price lines, P_1 and P_2, which represent two different price levels for paper (we take these prices as given for now, but will discuss what determines the price in a perfectly competitive market in the next part of the book).

The figure contains a lot of information, but let's take it step by step and start by reviewing costs. The starting point for the marginal cost and average variable cost curves is z, the price of raw materials. The average fixed cost is naturally very high at low quantities, which explains why the average total cost (ATC) is so high for low Q. As production increases, both marginal cost (MC) and average variable cost (AVC) rise. As explained in the previous chapter, this is because the marginal product of labour declines as production increases. At the same time, ATC falls because fixed costs are spread over more units. Continued production increases will eventually make the ATC curve rise again, pulled up by rising variable costs.

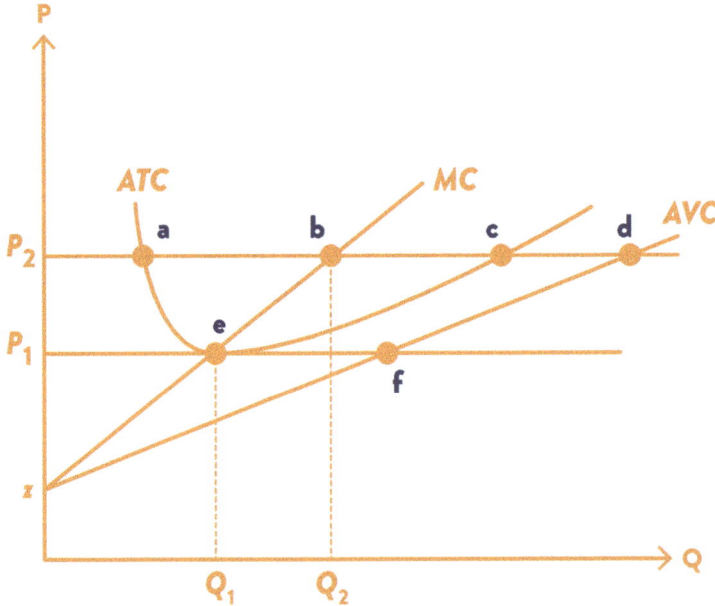

Fig. 7.1 Price, costs and profitability. *Note* The figure shows two price levels and the key cost functions: marginal cost (MC), average variable cost (AVC), and average total cost (ATC). Points a–f show different combinations of price and quantity that are relevant in an analysis of the firm's profitability

That was the cost side; now let's look at prices and profitability. We imagine the current market price for A4 paper is quite favourable, at P_2. At this price, there are four interesting production levels from a profitability perspective. Point a is where the average total cost (ATC) first crosses the price line. Point b is where marginal cost (MC) equals the price. Point c is where ATC crosses the price line for the second time. Point d is where the average variable cost (AVC) crosses the price line.

If we move along the P_2 line from the left, we first reach point a. Here, ATC equals the price, and profit is zero, while operating profits are positive since the price is higher than AVC at this production level. Continuing along the line, profit becomes positive as ATC falls below the price, leading us to point b, where marginal cost intersects the price. At this production level, Q_2, profit is maximised: the last unit produced costs exactly as much as it can be sold for. Since the price is also higher than ATC at this point, profit remains positive.

Further movement along the price line takes us to point c. We have moved away from the profit peak, so profit must have declined, and at point c, it is zero again. Here, the ATC curve crosses the price line a second time, driven by high variable costs. The final stop along the P_2 line is point d, where the AVC curve crosses the price line. At this point, operating profits are zero. This is because labour productivity has fallen so much that the income generated only covers labour and raw material costs, leaving nothing for the capital owners. Producing beyond point d leads to negative profitability even in the short run: the firm cannot cover its variable costs.

This was the situation for the higher price, P_2, but what if the price falls to P_1? A journey along the P_1 line has only two stopping points, e and f. The first is where marginal cost intersects the price line, representing the production level that maximises profit, Q_1. The last stop is f, where operating profits are zero.

Let's take a closer look at the profit-maximising choice at P_1, that is, point e. What is the profitability here? We observe that at this point, price equals ATC, meaning profit is zero. In other words, at this price, the highest profit the firm can achieve is zero profit. The capital owners are reasonably satisfied with this return, but it shouldn't get any lower! Therefore, price P_1 represents a long-run break-even point for the firm: if the price falls below this level and stays there, profit will be negative, and the owners will choose to sell the capital equipment and invest their money elsewhere for a better return, if they can. Remember that if costs are irreversible, meaning the capital equipment has no alternative use, then the money is effectively locked in the factory.

"This analysis provides some interesting insights," says Anna. "For example, it's quite possible that a firm should increase production in the short run but shut down in the long run or reduce production in the short run but invest in increased production capacity in the long run."

Conrad finds this confusing and asks for an explanation.

"Let me show you a similar figure to the one I just showed you," Anna says. She shows Conrad Fig. 7.2, with the same cost structure and again two prices: a high price, P_{high}, and a low price, P_{low} (which do not necessarily correspond to the high and low prices in the previous figure).

Imagine the price is low and the firm initially has production Q_a. What advice would you give to this firm? It should increase production to point b, but since ATC here is higher than the price, profit is negative, and in the long run, the capital owners will exit and the firm will shut down.

And what about if the price is high and the firm initially produces at Q_d? What advice should we give now? It should reduce production in the short run and choose point c, and since ATC here is lower than the price, positive profit is generated, making it attractive to invest more in the firm in the long run.

Conrad thinks this over and nods. Initially, he found these statements paradoxical, but with Fig. 7.2 and Anna's explanations, it makes sense. To be absolutely sure, he repeats the ideas in his own words.

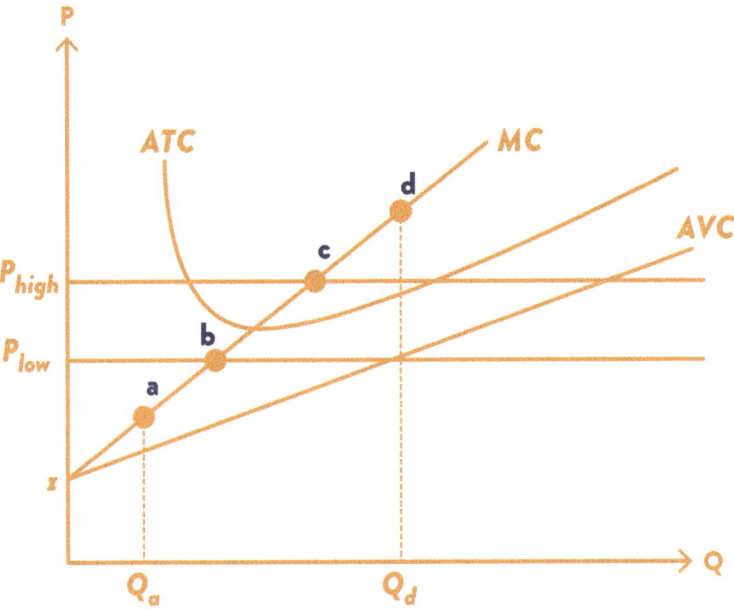

Fig. 7.2 Which production level should be chosen in the short and long run? *Note* The figure shows two price levels and the key cost functions: marginal cost (MC), average variable cost (AVC), and average total cost (ATC). With the low price P_{low}, the optimal production is at point b. A firm initially producing Q_a should therefore increase its production. However, since $ATC > P_{low}$ at point b, profit is negative, and in the long run, the firm will shut down. With the high price P_{high}, the optimal production is at point c. A firm initially producing Q_d should therefore reduce its production. Since $ATC < P_{high}$, at point c, profitability is very good, which means it is profitable to invest in the firm in the long run

Conrad: So, it can make sense to increase production in the short run, but shut down in the long run, because increasing production can lead to better utilisation of the capital that is already installed. However, even if you do the best that you can, it still won't be good enough when the price is so low: the profit will be negative regardless. And it can also make sense to reduce production today but expand in the future if today's production is very high relative to the installed capital, meaning production requires a lot of labour. Reducing output lowers costs and increases profit. And if the price is high enough, a pure profit is created, making it attractive to invest more in the factory in the future.

Anna: Yes, that's right. You've got it, Grandpa! And to answer your question about whether we should increase production in the paper mill, I'm happy to tell you the answer is yes!

Anna has checked the factory's production and the price of paper. She has found that the price is high and that today's production corresponds to point *b* in Fig. 7.2. But with the high price, production should be increased to point *c*. At the same time, she warns against getting too eager: if you go beyond point *c*, the factory might end up with lower profitability than it has today!

7.4   The Supply Curve

Based on the discussion above, we can now draw the firm's supply curve. The supply curve shows the firm's optimal choice of quantity for any given price, and as we have seen, this is determined by the firm's marginal costs.

Anna at a board meeting, lecturing about the profitability of the Mill

Anna attempts to engage the board of the paper mill with an analysis of the slope and level of the supply curve. The slope of the supply curve is influenced by the wage rate and the installed capital stock, as described in Math Box 7.1, where we found the marginal cost in the paper mill to be $MC = 2wQ/K_0 + z$. An increase in the wage level, w, raises the marginal cost for any quantity Q, resulting in a steeper supply curve. An increase in K_0 has the opposite effect, making the supply curve flatter. Why? Because more production capital makes labour more productive, meaning the firm needs fewer workers for any quantity produced, thus lowering marginal costs. The level of the supply curve is determined by the raw material cost, z. If raw material prices rise, the supply curve shifts upward; if they fall, it shifts downward.

The supply curve shows the firm's profit-maximising choice but says nothing about whether profit is positive or not. Anna knows the board cares about profit, so in Fig. 7.3 she also shows how the firm's profitability fares along the supply curve. Here, P_2 refers to the high price and P_1 to the low price from Fig. 7.1.

At the high price P_2, she recommends that the factory should produce a quantity Q_2, which generates a positive profit, $\pi > 0$. At the lower price P_1, the firm should produce the smaller quantity Q_1, and as we recall from Fig. 7.1, the profit here

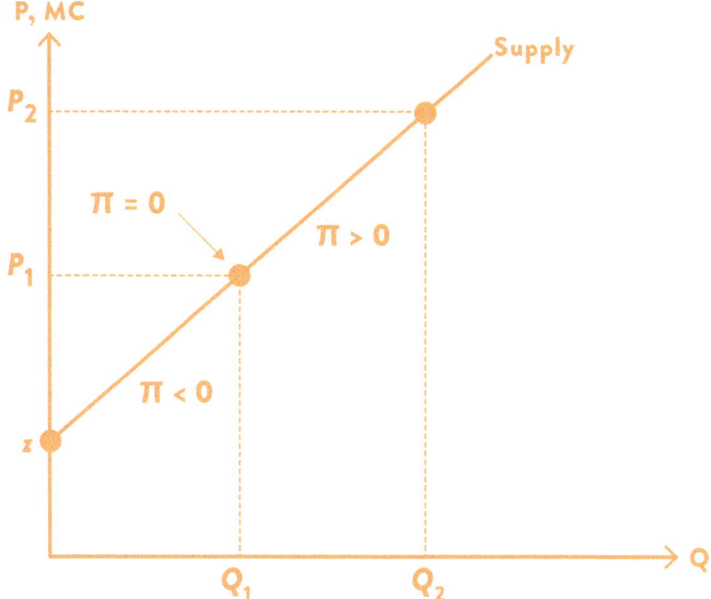

Fig. 7.3 The supply curve and profitability. *Note* The supply curve shows the relationship between price and the optimal quantity produced. When the firm takes the price as given, it maximises its profit by producing a quantity where marginal cost equals price. The supply is therefore determined by the marginal cost curve. Profitability varies along the supply curve: in the figure, price P_1 yields zero profit and thus marks the lowest price the firm can sustain in the long run. A price equal to z results in zero operating profit and marks the breaking point for production in the short run

is zero. To find the optimal quantity at an even lower price, we simply follow the marginal cost curve. We know that production at this level results in negative profit, while the operating profit is positive. Since a positive operating profit is better than none, Anna's recommendation is to continue production.

The lowest price the factory can accept even in the short run is z, which is the price of raw materials. Here the operating profit is zero, and any price below this means the firm cannot even cover its variable costs. The board nods approvingly at Anna's presentation (some nod because they are about to fall asleep), while expressing a hope for a future with good prices and high profits.

7.5   Demand for Labour

So far in this chapter, the focus has been on the firm's choice of production quantity. The supply curve is the most important result from this analysis: it shows the relationship between price and production that maximises profit. Higher price means higher production, according to the rule that price equals marginal cost.

To increase production in the short run, more labour is needed, so the product market and the labour market are naturally linked. For simplicity, let us disregard other variable inputs besides labour (that is, ignore raw materials and energy and set $Z = 0$), so that $VC = wL$. The marginal cost can then be written as:

$$MC = \frac{\partial VC}{\partial Q} = \frac{\partial (wL)}{\partial Q} = w\frac{\partial L}{\partial Q}$$

Note that:

$$\frac{\partial L}{\partial Q} = \frac{1}{\partial Q/\partial L} = \frac{1}{MP_L}$$

Marginal costs can therefore be expressed as:

$$MC = \frac{w}{MP_L}$$

We know that a profit-maximising firm chooses a quantity such that the marginal cost equals the price, which implies that:

$$P = MC \Rightarrow P = \frac{w}{MP_L} \Rightarrow PMP_L = w$$

This gives an expression for how many workers should be employed to maximise profit. If the number of employees is very high, the marginal product is low (it becomes crowded around the machines, see the discussion about technology in chapter 5.4), so the last worker hired produces little, and the firm loses money on that employment. The number of workers should be reduced. Conversely, if the number of employees is very low, productivity is very high (each worker has a lot of capital to work with), and the firm can increase profit by hiring more.

At the optimum, the last worker employed should create value for the firm exactly equal to the wage he or she receives. You can think of it as the workers needing to justify their wages.

Figure 7.4 illustrates the demand for labour, L^D, given by the value of the marginal product of labour (PMP_L). If the wage is w_a, the firm should choose point a, with a number of employees L_a. For $L < L_a$, $PMP_L > w$ and it pays to hire more. For $L > L_a$, $PMP_L < w$ and the last workers hired contribute less to value creation in the firm than they cost. A higher price P shifts L^D outwards in the diagram, while a lower price causes a shift inwards.

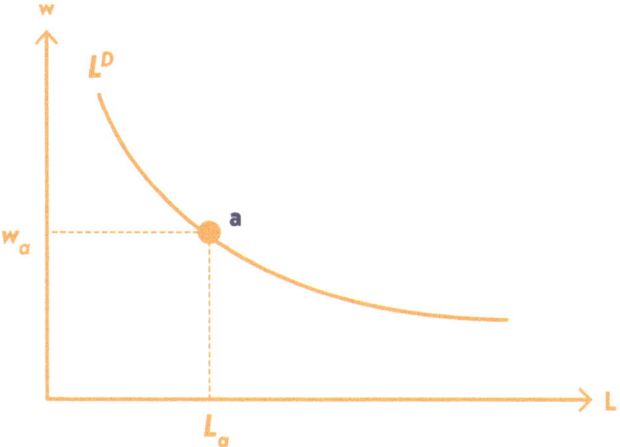

Fig. 7.4 Demand for Labour. *Note* The demand curve L^D shows the relationship between wages and the optimal number of workers in the firm, given by the value of the marginal product of labour (PMP_L). For example, at a wage w_a, the firm should employ L_a workers. Here, the value created by the last worker is exactly equal to the wage

7.6 Summary

Profit maximisation means that the firm chooses to produce a quantity such that marginal cost equals the market price. The marginal cost curve thus forms the firm's supply curve. But even if the firm does its best, that is not necessarily enough! If the price is sufficiently low, the firm should close down immediately. This happens when operating profits are negative, meaning that revenue is not enough to cover labour and raw material costs.

In the long run, it is also necessary that capital owners are satisfied with their investment: production must generate revenue that provides a return on capital at least as high as what they could get by placing their money elsewhere, such as in a bank. If the price is high enough, a pure profit (economic profit) will be created, which signals an incentive to invest more. On the other hand, a lower price that results in negative profit means that owners will try to sell off the capital equipment. In some cases, however, it is impossible to find buyers for the capital equipment—this is called an irreversible cost (sunk cost), and the relevant profitability measure in this case will be operating profits.

Finally, we looked at the firm's profit-maximising choice of the number of workers, derived from the first-order condition in the product market: price equals marginal cost. We found that the firm should hire workers as long as they create value which is at least as high as their wages. In optimum, the value of the last worker's production should be exactly equal to their wage.

7.7   Key Terms

Operating profit: Revenue minus variable costs.

Profit: Revenue minus total costs, which can also be expressed as operating profits minus fixed costs.

Pure profit (excess profit): When the firm generates income high enough to give the owners a return on capital that exceeds what they could have earned from the best alternative investment.

Supply curve: Shows the relationship between price and the quantity that maximises the firm's profitability. It is given by the firm's marginal cost curve.

7.8   Multiple-Choice Exercises

7.1: Operating profits: True or false?
Which of the following statements is true:

A. A positive operating profit means that the variable costs are covered
B. Variable costs can only be covered if the operating profit is maximised
C. In the short run, it is possible to operate profitably even if the operating profit is negative
D. A positive operating profit implies that the fixed costs are covered

7.2: Profit: True or False?
Which of the following statements is true:

A. A positive profit is necessary to cover the variable costs
B. Fixed costs can only be covered if the profit is maximised
C. In the short run, it is possible to operate profitably even if profit is negative
D. With zero profit the firm is not profitable even in the long run

7.3: Increase production?
When the price is higher than the marginal cost, one should:

A. Increase production, even if profit is negative
B. Increase production, but only if profit is positive
C. Keep production as it is, since the operating profit is positive
D. Reduce production to lower costs

7.4: Profit maximisation and profitability

Consider the figure below. The firm is initially producing at point a, with output Qa. Which of the following statements is correct?

A. Profit is maximised, but the firm is not profitable in the long run
B. Profit can be increased by increasing production, since the ATC curve is falling at point a
C. Profit is maximised, but the firm is not profitable even in the short run and should shut down immediately
D. Profit can be increased by reducing production, since this reduces variable costs

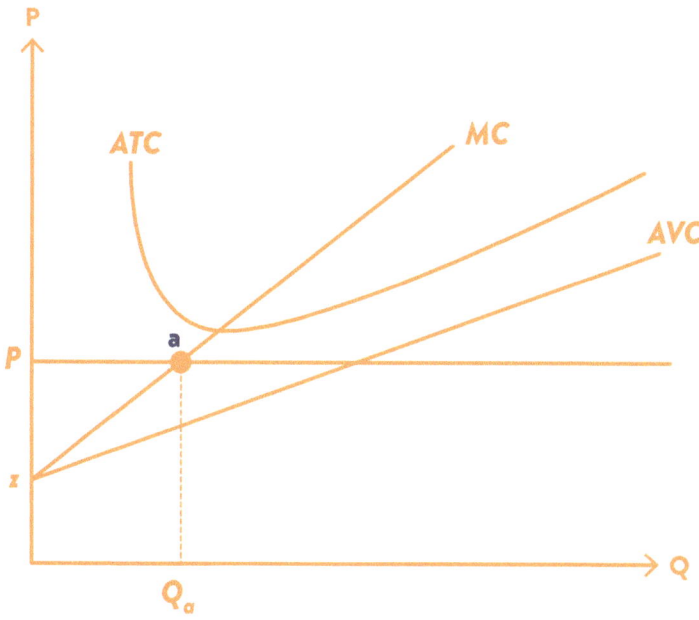

Solutions: 7.1 A; 7.2 C; 7.3 A; 7.4 A.

Part III

Market Theory

In this part of the book, we will study the perfectly competitive market, which describes a situation with many producers offering a homogeneous product, where all firms take the market price as given. We use the market for A4 paper as an example of such a market; another good example is electricity: many producers, identical product.

Chapter 8 shows how supply and demand form the market diagram that determines market equilibrium. We will examine how shocks to supply and demand affect this equilibrium, and how international trade influences the price and the quantities produced and consumed domestically.

In Chapter 9, we study market equilibrium from a normative perspective – that is, whether the outcome is good or bad. Using a criterion of economic efficiency, we show that the perfectly competitive market (under certain conditions) generates the best outcome for society. An important exception is situations involving externalities, such as environmental pollution, where the market should not be left to operate freely. We also consider the consequences of international trade and show that trade generates gains, but also creates winners and losers, which is why the topic is often politically contentious.

In Chapter 10, we examine how different types of policy affect Conrad's paper mill and all other paper mills in the country. The focus is on taxation, but we also study subsidies and direct market interventions in the form of price and quantity regulations. A thorough market analysis is necessary to understand how policy influences efficiency and distribution: the first impression is not always the right one!

Perfect Competition

8

Conrad notices that the price of paper varies from year to year, and wonders what's going on.

And then he's worried about international trade, and how it may cause problems for his factory—but Anna tries to reassure him that trade can also create new opportunities for growth!

8.1   Introduction

Conrad is worried about what increased international trade might mean for the factory

Conrad's paper mill faces intense competition from many other producers. Conrad may think his paper is the best, but in practice, customers can't tell the difference between A4 sheets from different suppliers: it's a homogeneous product. We'll use paper as an example of a perfectly competitive market, characterised by many firms producing identical goods and all taking the price as given. This contrasts with markets characterised by imperfect competition—the topic of the final part of the book—where one or a few producers recognise that their decisions can influence the market price.

We begin this chapter by deriving the market supply and demand curves. We then combine them to find the market equilibrium—the point where supply meets demand—and explore how this equilibrium changes in response to shocks on the supply or demand side. This analysis will help Conrad understand why the price of paper can fluctuate from year to year. We'll see that sometimes a higher price is accompanied by lower production; other times, both price and output increase; and in some cases, the price rises without much change in output at all. It all depends on the type of shock, and the market model offers a useful framework for thinking through these complex relationships.

Finally, we turn to international trade. Is trade necessarily bad news for Conrad, or could it, as Anna points out, open new opportunities for the paper mill?

But before we dive into these important questions, we need to familiar ourselves with the theoretical tools. In the first part of the book, we developed consumer theory, and in the second, producer theory. There, the focus was on individual consumers and firms. Now, we shift our attention to the market as a whole—the combined behaviour of all consumers and producers. We begin by aggregating demand and supply and then present the market model. This gives us the market equilibrium—and that's our starting point for analysing a range of shocks. Let's see what happens!

8.2   Aggregating Demand

The first part of this book focused on consumers. A key theme was how changes in prices and income influence our choices. Apart from very special cases (see Giffen goods), an increase in the price of a good leads to a decrease in the quantity demanded, while a price reduction leads to an increase in demand. We take this normal case as our starting point and assume that individual demand (indicated with a superscript i) is described by a linear function:

$$Q_i^D = \alpha_i - \beta_i P \tag{8.1}$$

Here, α_i (the Greek letter alpha) is a variable that shifts individual demand, due to a change in income or the price of another good, while β_i (the Greek letter beta) indicates how strongly demand responds to a change in price—for example, depending on the availability of close substitutes.

The point where the demand curve intersects the vertical axis represents the maximum amount the consumer is willing to pay for the good in question, often referred to as the consumer's choke price. At this price, or any price higher, demand is reduced to zero.

$$Q_i^D = 0 \Rightarrow P_{D_i}^{choke} = \frac{\alpha_i}{\beta_i} \quad \text{The consumer's choke price} \tag{8.2}$$

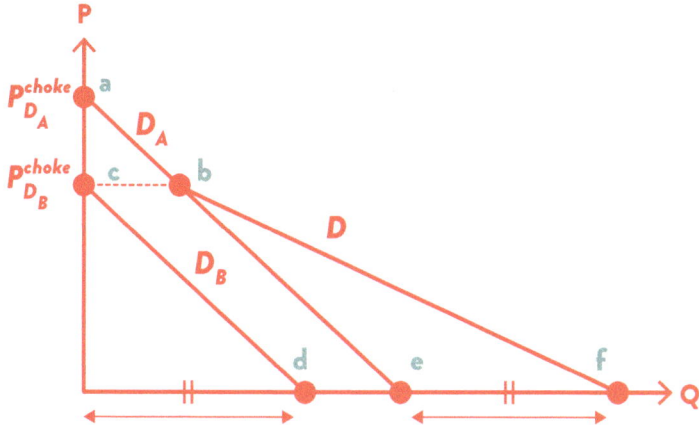

Fig. 8.1 Aggregating demand. *Note* Anna's demand curve is labelled D_A and Brian's is D_B. The two curves have the same slope, meaning $\beta_A = \beta_B$, but Anna's curve intersects the horizontal axis further out than Brian's, which means that $\alpha_A > \alpha_B$. Anna's choke price is $P_{D_A}^{choke}$, shown at point a, while Brian's choke price $P_{D_B}^{choke}$ is lower, shown at point c. The market demand curve D is found by horizontally summing the two individual demand curves. For prices above point c, only Anna demands the good, so the market demand equals D_A. Brian only enters the market when the price falls below his choke price, which happens at point b, and where there is therefore a kink in the market demand curve. The combined demand curve intersects the horizontal axis at point f, which is the sum of the maximum quantities that Anna and Brian would consume

Imagine a market with two consumers, Anna and Brian, with demand functions D_A and D_B as shown in Fig. 8.1. Anna's choke price ($P_{D_A}^{choke}$) is given at point a, while Brian's choke price ($P_{D_B}^{choke}$) is shown at point c. The demand curves intersect the horizontal axis where $Q = \alpha_i$. This is a consequence of the linear function and implies that there is a limit to how much one would want to consume, even at a zero price. In the figure, the two consumers' demand curves are parallel, meaning they have the same price sensitivity: $\beta_A = \beta_B$.

The market demand, D, where the market in this case consists of Anna and Brian, is found by horizontally summing their individual demand curves. At prices above Brian's choke price, only Anna demands the good, so market demand is given by $D = D_A$. When the price falls below this level, Brian also enters the market. As a result, the market demand curve D has a kink at point b, and for prices below this point, market demand is given by $D = D_A + D_B$.

We can find the point where total demand intersects the horizontal axis by summing the maximum quantities demanded by the two individuals—for Anna, this is point e, and for Brian, point d, giving a total at point f.

8.3   The Market Demand Curve, Its Level and Slope

Since markets typically consist of many consumers, we often ignore such kinks
and assume that the market demand curve is linear:

$$Q^D = \alpha - \beta P \quad \text{Market demand} \tag{8.3}$$

We will now take a closer look at the demand curve: What determines its level and
slope? This is important for our later analysis of how demand-side shocks affect
the economy.

In the market demand function, the variable α determines the level of the
demand curve, while β determines its slope. For example, higher income (assuming the good is normal, as discussed in the consumer theory in Part 1 of the book)
leads to an increase in α, shifting the demand curve from D_1 to D_2 in Fig. 8.2.
An increase in β makes the curve flatter, as shown by D_3. Demand is now more
sensitive to price—possibly because a substitute for Q has entered the market.

One can also imagine a combination of both: an increase in both α and β. A
relevant example is growth in electricity demand over time. The shift to electric
vehicles has pushed demand outward (higher α), while investments in heat pumps,
solar panels, and batteries have enabled people to reduce electricity consumption
when prices are high—in other words, demand has become more price-sensitive
(higher β).

As this example illustrates, the time dimension can influence consumption. This
is something we also saw in producer theory (see especially Chapter 6): in the short
run, producers must work with a given stock of installed capital, whereas in the

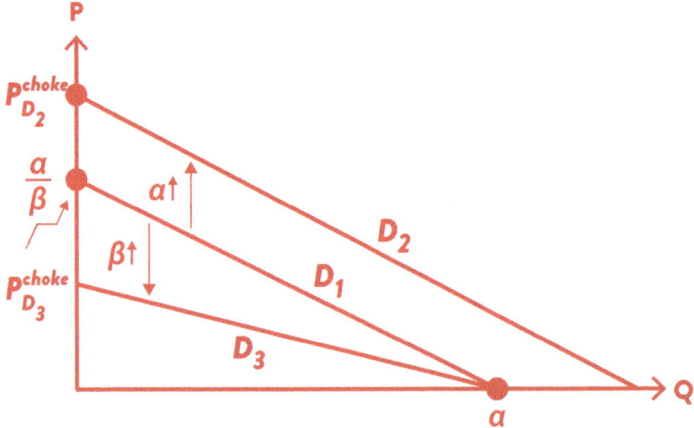

Fig. 8.2 The Level and Slope of the Demand Curve. *Note* An increase in α shifts the demand curve
upward, from D_1 to D_2, while an increase in β makes the demand curve flatter, from D_1 to D_3. Note
that a change in β does not affect the point where the demand curve intersects the horizontal axis,
which is given by $P = 0 \Rightarrow Q^D = \alpha$

long run, they can adjust their capital stock. Similarly, consumers may have greater opportunities to adjust their behaviour over the long run, allowing them to benefit more from low prices and protect themselves against high prices.

The price sensitivity of demand is often expressed as an elasticity—that is, as a percentage change—which provides a convenient metric for measuring the size of the response:

$$\varepsilon^D = -\frac{\partial Q^D}{\partial P}\frac{P}{Q^D} \qquad \text{Demand elasticity} \qquad (8.4)$$

Here, I have chosen to include a minus sign on the right-hand side so that the elasticity becomes a positive number (since $\partial Q^D/\partial P < 0$). More on the elasticity of a linear demand function can be found in Math Box 8.1.

Math Box 8.1 Elasticity of demand

Suppose we have the demand function:

$$Q^D = \alpha - \beta P$$

If we differentiate quantity with respect to price, we find that:

$$\frac{\partial Q^D}{\partial P} = -\beta$$

Note that we can write the demand function in its inverse form, that is, with price expressed as a function of quantity, as follows:

$$P = \frac{\alpha - Q^D}{\beta}$$

We then substitute the inverse demand function and the result of the differentiation $\partial Q^D/\partial P$ into the definition of the price elasticity of demand, and find:

$$\varepsilon^D = -\frac{\partial Q^D}{\partial P}\frac{P}{Q^D} = -(-)\beta\frac{\alpha - Q^D}{\beta Q^D} = \frac{\alpha - Q^D}{Q^D}$$

We note that the elasticity depends on Q, that is, on where we are along the demand curve. If $\varepsilon^D = 1$, meaning that a one percent increase in price leads to a one percent decrease in quantity demanded, we say that demand is unit elastic. We can find the point on the linear demand curve where demand is unit elastic as follows:

$$\varepsilon^D = 1 \Rightarrow \frac{\alpha - Q^D}{Q^D} = 1 \Rightarrow Q^D = 0.5\alpha \text{ Unit elastic demand}$$

If $\varepsilon^D > 1$, meaning that a one percent increase in price leads to a reduction in demand greater than one percent, we say that demand is elastic. We can find the region on the linear demand curve where demand is elastic as follows:

$$\varepsilon^D > 1 \Rightarrow \frac{\alpha - Q^D}{Q^D} > 1 \Rightarrow Q^D < 0.5\alpha \text{ Elastic demand}$$

And finally, if $\varepsilon^D < 1$, meaning that a one percent increase in price leads to a reduction in demand smaller than one percent, we say that demand is inelastic. The region on the linear demand curve where demand is inelastic is:

$$\varepsilon^D < 1 \Rightarrow \frac{\alpha - Q^D}{Q^D} < 1 \Rightarrow Q^D > 0.5\alpha \text{ Inelastic demand}$$

Figure 8.3 illustrates an important insight from Math Box 8.1: the price elasticity of demand varies along a linear demand curve. Halfway to the point where the demand curve intersects the horizontal axis, demand is unit elastic—this is at point a, where quantity $Q = 0.5\alpha$. Demand is inelastic to the right of this point, and elastic to the left.

With a linear demand curve, a given increase in price results in the same absolute reduction in quantity demanded along the entire curve. However, this means the percentage change in quantity demanded varies depending on the starting point: the percentage reduction will be small (inelastic) when demand is initially high, and large (elastic) when the initial quantity demanded is low.

As a contrast to the normal case of a downward-sloping demand curve, it can be useful to consider two special cases, where demand is either horizontal or vertical, as shown in Fig. 8.4. A horizontal demand curve means that demand is extremely price-sensitive, so that even a small increase in price causes the entire demand to disappear.

$$\frac{\partial Q^D}{\partial P} = -\infty \Rightarrow \varepsilon^D = \infty \text{ Horisontal demand curve}$$

This is often referred to as perfectly elastic demand. The other extreme is a vertical demand curve, which means that demand is completely insensitive to price

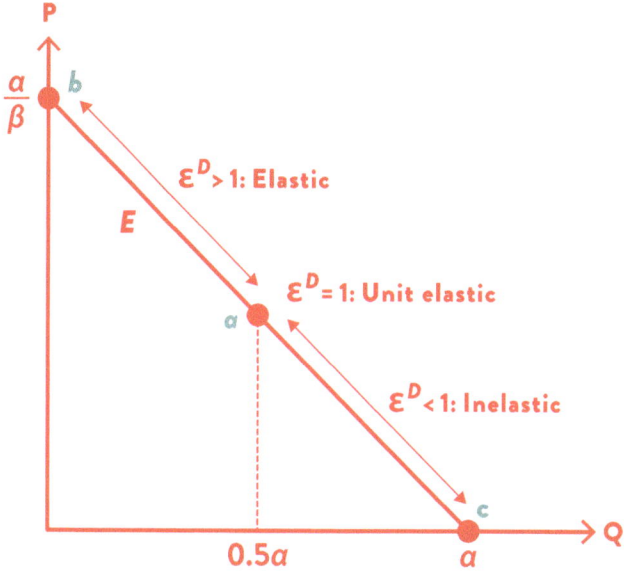

Fig. 8.3 Price Elasticities of Demand. *Note* The figure shows how the price elasticity of demand varies along a linear demand curve. At $Q = 0.5\alpha$, demand is unit elastic, $\varepsilon^D=1$, meaning a one percent increase in price leads to a one percent decrease in quantity demanded. For $Q < 0.5\alpha$, demand is elastic, $\varepsilon^D > 1$, meaning a one percent increase in price reduces demand by more than one percent. For $Q > 0.5\alpha$, demand is inelastic, $\varepsilon^D < 1$, meaning a one percent increase in price reduces demand by less than one percent

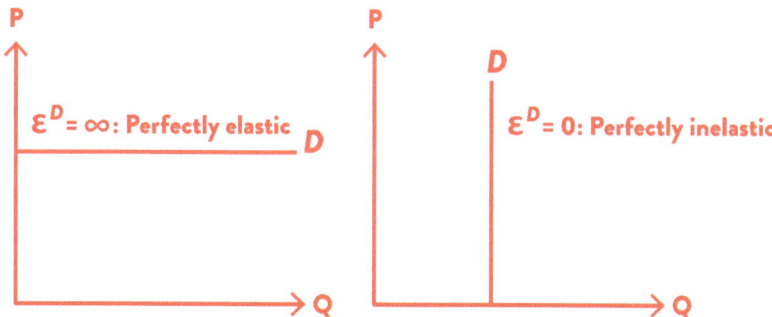

Fig. 8.4   a Perfectly elastic demand **b** Perfectly inelastic demand. *Note* A horizontal demand curve (8.4a) means demand is perfectly elastic: a small decrease in price leads to an infinitely large increase in quantity demanded, while a small increase in price causes no one to want to buy the good. A vertical demand curve (8.4b) means demand is perfectly inelastic: consumers are willing to pay any price to purchase a given quantity

changes: consumers want a fixed quantity, no matter the cost.

$$\frac{\partial Q^D}{\partial P} = 0 \Rightarrow \varepsilon^D = 0 \quad \text{Vertical demand curve}$$

This is called perfectly inelastic demand.

Generally, we will work with a downward-sloping demand curve, but both in this main textbook and in the Workbook, we will also use examples with either perfectly elastic or perfectly inelastic demand to highlight important theoretical points related to price sensitivity in demand.

8.4   Aggregating Supply

You may recall from producer theory (Part 2 of the book) that a producer (who takes the price as given) chooses a quantity such that marginal cost equals price. A higher price therefore makes it profitable to increase production. As a simplification of this relationship, assume that a firm's supply (where the subscript i denotes an individual firm) is given by the linear function:

$$Q_i^S = -\gamma_i + \delta_i P \quad \text{The firm's supply function} \tag{8.5}$$

Here, γ_i (the Greek letter gamma) is a variable that determines the level of supply, while δ_i (the Greek letter delta) determines the slope of the supply curve. A producer's choke price—the price at which supply drops to zero—is given by:

$$Q_i^S = 0 \Rightarrow P_{S_i}^{choke} = \frac{\gamma_i}{\delta_i} \quad \text{Producer choke price} \tag{8.6}$$

To find the market supply curve, we sum the supply curves of all firms. Imagine the market consists of two perfectly competitive producers, Conrad and his colleague—you know, the one with the modern paper mill Conrad dreamed of (see the introduction to Chapter 6). Figure 8.5 shows Conrad's supply curve as S_C while his colleague, with his production unit up north, has the supply curve S_N.

There are three interesting points on Conrad's supply curve, marked a, b, and c. First, notice that S_C crosses the horizontal axis at a negative quantity, point a. We can see this by setting $P = 0$ in Eq. (8.5), which gives $Q_C^S = -\gamma_C$. Since negative production is impossible, the supply curve is dashed up to Conrad's choke price, $P_{S_C}^{choke}$, marked at point b. Only when the price exceeds $P_{S_C}^{choke}$ does it become profitable for Conrad to supply paper to the market. For a market price P_f, for example, he wants to produce a quantity Q_c, marked at point c on S_C.

The supply curve from the producer up north, S_N, lies below Conrad's, we see that it starts at the origin (point d) and has the same slope as S_C. What does this tell us about the parameters in the competitor's supply function? Since it starts at the origin, it must mean that $\gamma_N = 0$, so $P_{S_N}^{choke} = 0$. And since the two supply curves have the same slope, it must mean that $\delta_C = \delta_N$. The fact that, in the

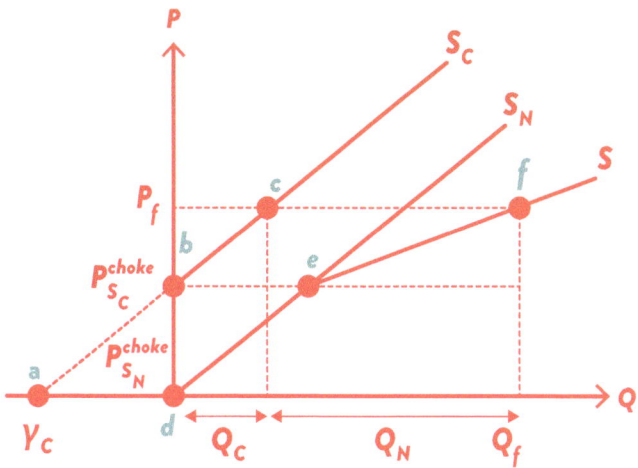

Fig. 8.5 Aggregating supply. *Note* Conrad's supply curve S_C hypothetically starts at point *a*, but he must have a price at least as high as his choke price, $P_{S_C}^{choke}$, given by point *b*, for production to be profitable. The supply curve of the producer in the North, S_N, starts at the origin, meaning $\gamma_N = 0$. The two supply curves have the same slope, so $\delta_C = \delta_N$. The market supply curve S is found by horizontally summing the two firms' supply curves. Conrad will only supply a positive quantity if the price exceeds his choke price $P_{S_C}^{choke}$. Thus, up to point *e*, the market supply is given by S_N alone. For prices higher than this, Conrad also enters the market, and at a price P_f, the market supply is given by point *f*, with total quantity Q_f, where Conrad supplies Q_C and the producer in the North supplies Q_N

figure, $\gamma_N = 0$ can be interpreted as the factory in the North operating more efficiently—it has lower costs related to inputs and energy than Conrad's factory, and can make a profit selling paper even if the price approaches zero.

The total market supply, S, is the horizontal sum of the two firms' supply curves. For prices $P < P_{S_C}^{choke}$, only the factory in the North supplies paper, so here the market supply is given by S_N. But when the price rises above this critical level, Conrad enters the market, so from point *e* upwards, we must add his supply. For example, at price P_f, the two factories together supply a quantity Q_f. We see that Conrad's factory supplies less to the market than the factory in the North, $Q_C < Q_N$, and this difference is due to Conrad's higher marginal costs.

8.5 The Market Supply Curve, Its Level and Slope

Similar to the demand side, when we sum the supply from all paper factories, we simplify by assuming a linear supply curve without kink points:

$$Q^S = -\gamma + \delta P \qquad \text{The market supply curve} \qquad (8.7)$$

The same factors that influence the slope of an individual firm's supply curve will also affect the market supply curve. The variable γ determines the level of the supply curve. An increase in the price of raw materials and energy causes γ to rise, shifting the supply curve upward, as shown in Fig. 8.6 by the shift from S_1 to S_2. A change in δ affects the slope of the supply function. For example, lower wages or more installed production capital increase δ, making the supply curve flatter (see the discussion of this in Chapter 7), as illustrated by S_3 in Fig. 8.6.

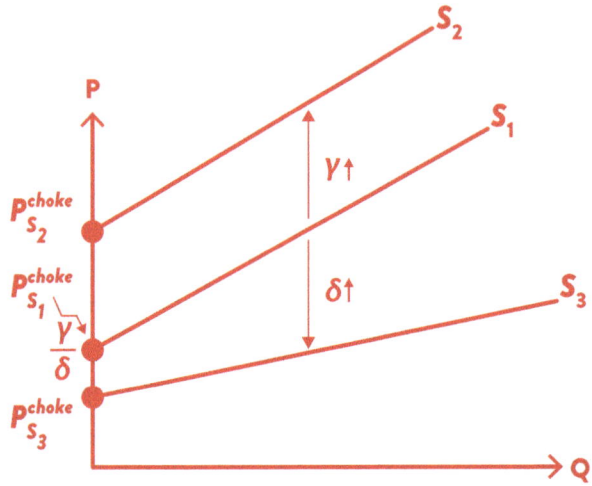

Fig. 8.6 The Level and Slope of the Supply Curve. *Note* An increase in γ shifts the supply curve upward, from S_1 to S_2, while an increase in δ makes the supply curve flatter, from S_1 to S_3. Note that a change in δ also affects the point where the supply curve intersects the vertical axis, that is, the producers' choke price, since this is given by $P_S^{choke} = \gamma/\delta$

The price sensitivity in supply can be expressed as an elasticity:

$$\varepsilon^S = \frac{\partial Q^S}{\partial P} \frac{P}{Q^S} \qquad \text{Supply elasticity} \tag{8.8}$$

How much does supply increase, measured in percent, when the price increases by one percent? A high elasticity means supply responds strongly to price changes, i.e., the supply curve is relatively flat, while a low elasticity means low price sensitivity and a steeper supply curve.

In Math Box 8.2, we derive the supply elasticity for a linear supply function.

Math Box 8.2 Elasticity of Supply
Assume that supply is given by:

$$Q^S = -\gamma + \delta P$$

If we differentiate quantity with respect to price, we find that:

$$\frac{\partial Q^S}{\partial P} = \delta$$

The next step is to rewrite the supply function in inverse form, that is, as price as a function of quantity (this can be interpreted as the marginal cost curve):

$$P = \frac{\gamma + Q^S}{\delta}$$

We insert this expression for the inverse supply function, along with the result from the derivative $\partial Q^S / \partial P$, into the definition of the supply elasticity and find:

$$\varepsilon^S = \frac{\partial Q^S}{\partial P}\frac{P}{Q^S} = \delta\frac{\gamma + Q^S}{\delta Q^S} = \frac{\gamma + Q^S}{Q^S}$$

We see that for $\gamma > 0$, the supply elasticity $\varepsilon^S > 1$, while $\gamma = 0$ gives $\varepsilon^S = 1$. Note that, in contrast to the price elasticity of demand, we do not need to place a minus sign in front of the elasticity definition to get a positive value, since $\partial Q^S / \partial P > 0$.

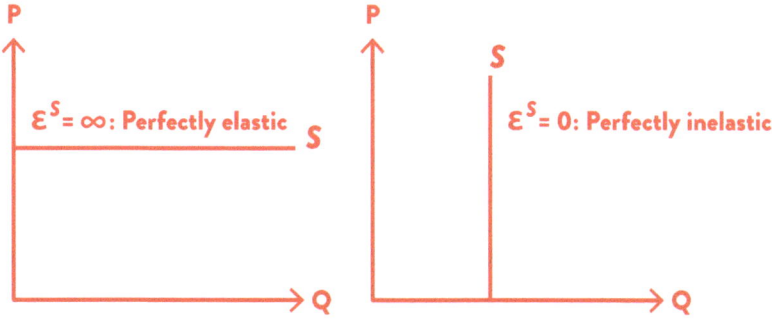

Fig. 8.7 **a** Perfectly Elastic Supply. **b** Perfectly Inelastic Supply. *Note* A horizontal supply curve (8.7a) means that supply is perfectly elastic: a small increase in price leads to an infinitely large increase in the quantity supplied, and similarly, a small reduction in price causes producers to stop supplying the product entirely. A vertical supply curve (8.7b) means that supply is perfectly inelastic: firms are willing to supply a fixed quantity regardless of the price—we can think of it as a capacity limit

As a contrast to the normal case of an upward-sloping supply curve, consider the extreme cases shown in Fig. 8.7. A horizontal supply curve implies that supply is extremely price-sensitive, even a small reduction in price causes the entire supply to disappear:

$$\frac{\partial Q^S}{\partial P} = \infty \Rightarrow \varepsilon^S = \infty \quad \text{Horisontal supply curve}$$

This is perfectly elastic supply. A vertical supply curve means that supply is completely unresponsive to price changes, producers cannot supply more than this. Think of it as a capacity limit:

$$\frac{\partial Q^S}{\partial P} = 0 \Rightarrow \varepsilon^S = 0 \quad \text{Vertical supply curve}$$

The supply is then perfectly inelastic.

Throughout, we typically use an upward-sloping supply curve as shown in Fig. 8.6, but we will also present examples of situations with either perfectly elastic or perfectly inelastic supply, to illustrate the importance of elasticities for market outcomes. This applies, for instance, to electricity production, where different technologies make the supply curve take on a step-like shape, with perfectly elastic supply on each step, and perfectly inelastic supply between steps (see Sect. 6.3 on producers hitting the wall). More on this shortly, but we begin with a standard market equilibrium diagram.

8.6   Market Equilibrium

Conrad is concerned about the price of paper and has noticed that it has fluctuated over recent years. He wants to understand what might explain these swings and asks Anna for an analysis.

Anna combines supply and demand and presents the market model to him using Fig. 8.8. The demand curve is D and the supply curve is S, and at the intersection point, point a, we have a market equilibrium with price P_a. Math Box 8.3 shows the derivation of the market price using linear supply and demand functions.

When we say that point a in Fig. 8.8 is an equilibrium, it means the market is balanced here, and that for any other price than P_a, there will be forces pushing the economy back toward this balance point.

For example, consider the price P_b, which is higher than P_a. At this price, supply S is greater than demand D: the quantity producers offer exceeds the quantity consumers are willing to buy. In other words, there is a surplus supply in the market, which will cause the price to fall. This makes sense: if there is much more paper available than buyers, producers will be expected to lower their prices to sell their products. This dynamic will stop only when the equilibrium price P_a is reached.

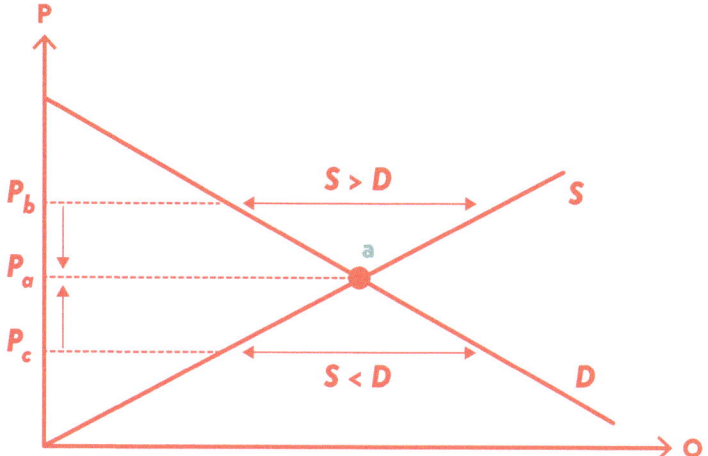

Fig. 8.8 Market Equilibrium. *Note* The market equilibrium is given by the price that makes supply equal to demand, S = D. In the figure, this is at price P_a. A price higher than the equilibrium price, such as P_b, means there is a surplus supply in the market, S > D. We then expect the price to fall. Similarly, a price lower than the equilibrium price, such as P_c, will lead to excess demand, D > S. The price will then rise

Similarly, if we start with a price lower than P_a, for example P_c, we have a situation where demand D is greater than supply S, meaning excess demand. Many consumers want to buy paper but cannot. Consumers will then start bidding against each other, pushing prices up, and this process stops only when the price reaches P_a.

In this way, P_a represents a stable point where supply and demand balance, at least as long as the supply and demand curves remain unchanged. The most interesting aspect of the market model, however, is that it helps us analyse what happens when supply and demand do not stay constant—that is, when so-called shocks occur in the market.

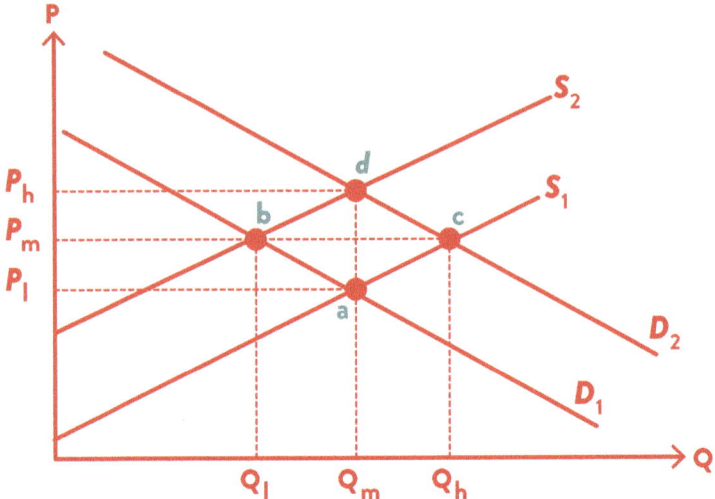

Fig. 8.9 Shifts in demand and supply. *Note* Initially, demand is given by D_1 and supply by S_1, and the market equilibrium is at point a, with price P_l and quantity Q_m which we can think of as a "low" price and "medium" quantity. A negative supply shock, for example due to an increase in raw material prices, causes the supply curve to shift upward, from S_1 to S_2. The new equilibrium is at point b, where the equilibrium price rises from low to medium, from P_l to P_m, and quantity falls from medium to low, from Q_m to Q_l. A positive demand shock, for example due to higher income or an increased price of a substitute, causes the demand curve to shift upward, from D_1 to D_2. The new equilibrium is at point c, on the original supply curve S_1, where the equilibrium price rises from low P_l to medium P_m, and quantity rises from medium Q_m to high Q_h. A combination of the two shocks leads to an equilibrium at point d, with a high price, P_h but unchanged quantity, Q_m

---

Math Box 8.3 Equilibrium price

Let demand be given by $Q^D = \alpha - \beta P$ and supply by $Q^S = -\gamma + \delta P$. In equilibrium $Q^D = Q^S$, which implies that:

$$\alpha - \beta P = -\gamma + \delta P$$

The equilibrium price is therefore:

$$P^* = \frac{\alpha + \gamma}{\beta + \delta} \quad \text{Equilibrium price}$$

We see that the shift parameters α and γ are in the numerator, while the slope parameters β and δ are in the denominator. An increase in price can thus be caused by an increase in α, which shifts the demand curve positively; an increase in γ, which shifts the supply curve negatively; a reduction in β (less price-sensitive demand), which flattens the demand curve; or a reduction in δ (less price-sensitive supply), which makes the supply curve steeper.

Anna explains that there are two possible reasons for changes in the price of paper: one is due to shocks on the supply side, and the other due to shocks on the demand side. She uses Fig. 8.9 to illustrate.

"Imagine," she says, "that we start at point a, with a relatively low price P_l and a medium quantity Q_m. Now suppose there is a shock on the supply side that shifts the supply curve—in the figure, it shifts to the left. The new price becomes P_m, a medium-high price."

Conrad asks what she means by a shock and what could explain the shift in the supply curve. Anna explains that supply shocks include anything that changes production costs for a given quantity. This could be changes in the price of electricity or pulp, which in the figure are captured by changes in the shift parameter γ. "This year electricity is very expensive," she says, "which causes γ to increase and shifts the supply curve from S_1 to S_2, resulting in a new equilibrium at point b, where the price has increased from low to medium. If electricity became cheaper, γ would decrease and the supply curve would shift in the opposite direction."

"I see that the quantity produced has fallen in the figure," says Conrad, "which fits with what I'm experiencing—it's harder to sell paper now than before. But does a higher price for paper always mean higher costs and lower production?"

Anna explains that this is not necessarily the case, pointing to demand shocks instead. "Imagine schools getting tired of screens and the government recommending the purchase of real books instead of digital learning materials. Books require paper, which increases α, causing a positive shift in the demand curve from D_1 to D_2. We see that the new equilibrium is at point c, with a higher price (rising from low to medium, P_m) and a high quantity Q_h.

Conrad points to point d in the figure. "Here I see the price of paper has risen a lot, and it seems to be due to a combination of higher electricity prices and increased demand for paper."

"That's right, Grandpa," Anna says, "and notice that the quantity produced doesn't change much. This is because two opposing forces are at work. On one hand, the demand for more schoolbooks pushes demand up, raising both price and quantity, a movement from a to c along the supply curve. On the other hand, higher electricity prices reduce supply, causing a movement along the demand curve D_2 from point c to d."

8.7   Stepwise Supply

So far, the discussion has been based on a linear supply function, but in Part 2 of the book, on producer theory, we discussed different types of technologies. We looked at technologies where the installed capital defines an absolute capacity limit for production. This applies when the production factors are complementary—that is, Leontief production. In such cases, producers can hit a wall, in the sense that it's not possible to increase production beyond a certain level.

A relevant example is electricity production, which requires large investments in dams and turbines. The complementary input is the energy from water, wind, gas, or coal. Water and wind, which we can refer to as *green* energy, cost nothing to produce—you can think of it as manna from heaven. The green producer can hit the wall (denoted by quantity Q_G) either because the manna stops falling (little rainfall, wind, or sunshine) or because installed capacity limits output (e.g., the reservoirs are full, all windmills are at full capacity, the number of solar panels installed, etc.).

The *brown* producer has access to gas and coal at a given unit price z and hits the wall (at Q_B) when the capacity of the production facility is reached. The marginal costs, and thus the supply functions, of the two types of producers are illustrated in Fig. 8.10a: S_G for the green and S_B for the brown producer. For the green producer, energy is free, so the horizontal part of the supply curve is constant, at zero. The vertical part can move left or right depending on access to water, wind, or sun, though the installed capital (e.g., the dam, the windmills or the solar panels) sets an upper limit. For the brown producer, the horizontal part of the supply curve depends on the price of coal and gas, while the vertical part is unchanged in the short run, determined by installed production capacity.

We note that for each technology, marginal cost is constant up to the capacity limit and then becomes vertical. A horizontal supply curve indicates perfectly elastic supply, while the vertical part represents perfectly inelastic supply.

If we now add a demand curve D_1, we can find the market equilibrium, shown as point a in Fig. 8.11. Think of this as summer, when there is no need for heating and demand for electricity is low. All demand is covered by green energy, and the capacity constraint is not binding. The marginal cost of production, and therefore the price, is zero.

What happens when winter arrives and demand shifts to D_2? The new equilibrium is at point b, with an equilibrium price P_b. Note that the price here is high even though the electricity is still entirely produced using green energy. Why is

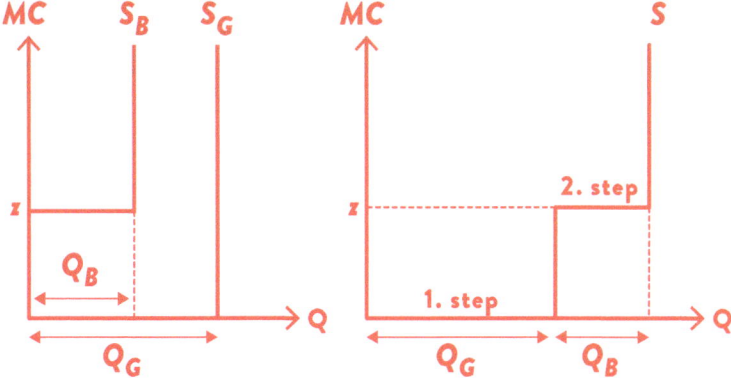

Fig. 8.10 **a** Supply with different technologies. **b** Stepwise supply. *Note* Fig. 8.10a shows a situation where a product, electricity, can be produced using two different technologies: green and brown. Green energy uses water and wind as input factors, which we assume are free. Maximum production here is Q_G, determined either by the availability of water, wind or sun, or by the size of the dams or number of wind turbines and solar panels. Brown energy uses gas and coal as inputs, with marginal cost z, and installed capital that sets a production ceiling at Q_B. Figure 8.10b shows the market supply curve with the two technologies, and we see that it takes a stepwise form. Up to the capacity limit for green energy, marginal cost is zero (first step), while beyond this, up to the capacity limit Q_B, the marginal cost is defined by the cost of gas and coal, z (second step)

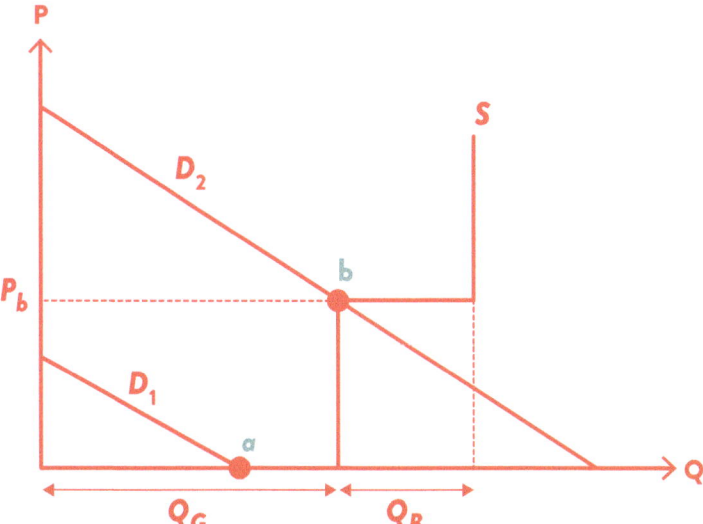

Fig. 8.11 Market equilibrium with stepwise supply. *Note* This figure builds on Fig. 8.10b by adding the demand side. With low demand D_1, there is more than enough capacity in green energy to meet the entire demand, so the marginal cost, and therefore the price, is zero. If demand increases to D_2, all available capacity in green energy is used, and the market price rises to P_b to balance supply and demand. The entire demand is still covered by green energy, but if demand increases further, brown energy must also be used

the electricity price so high when rain is pennies from heaven and the sun and wind are also gifts of nature? The key is that the market price balances supply and demand. The high price is necessary to reduce demand to the level that producers can deliver, in this case, Q_G.

If demand shifts even further outward—say, on an extra cold winter day—then brown energy must also be used to meet demand. The marginal cost of brown energy will then determine the market price, represented in the figure by the second step of the supply curve. That will be the price of electricity until the brown producer also hits the capacity limit.

8.8   International Trade

Conrad is worried. The World Trade Organization is working to promote trade liberalisation and has turned its attention to the market for paper. How will increased international trade affect local paper production? Assume that without such an agreement, there is no international trade in paper at all. This is called autarky: each country is self-sufficient in the good in question.

Under free trade, however, the international price of A4 paper, P_{int}, determines the market price at home as well. Domestic consumers can buy paper from anywhere at this price, and local producers can sell anywhere at the same price. The effect of international trade on the local market depends on how P_{int} compares with the autarky price, P_{aut}. Note that in the special case where $P_{int} = P_{aut}$, there will be no international trade in this good at all.

Math Box 8.4 International trade
How does the international price affect exports and imports? Assume that demand is given by $Q^D = \alpha - \beta P$ and supply by $Q^S = -\gamma + \delta P$. With international trade, the price at home is determined by the international price, $P = P_{int}$. Exports are defined as $Q_{exp} = Q^S - Q^D$, that is, the quantity produced that exceeds domestic demand. Using the supply and demand functions and the fact that with trade, $P = P_{int}$, we can write exports as:

$$Q_{exp} = -\gamma + \delta P_{int} - (\alpha - \beta P_{int})$$

This can be written as:

$$Q_{exp} = -(\alpha + \gamma) + (\beta + \delta)P_{int}$$

We multiply the first term on the right-hand side by $\beta + \delta$ in both the numerator and the denominator, simplify, and get:

$$Q_{exp} = (\beta + \delta)\left(P_{int} - \frac{\alpha + \gamma}{\beta + \delta}\right)$$

The last term in the expression above looks familiar! From Math Box 8.3, we know that this is the same as the equilibrium price in a market without trade—what we here refer to as the autarky price, P_{aut}. We can therefore write:

$$Q_{exp} = (\beta + \delta)(P_{int} - P_{aut}) \quad \text{Exports}$$

And since imports are $Q_{imp} = Q^D - Q^S$, (you can think about imports as negative exports), we have

$$Q_{imp} = (\beta + \delta)\left(P_{aut} - P_{imp}\right) \quad \text{Imports}$$

From these expressions for exports and imports, we can draw two key conclusions. First, when $P_{int} > P_{aut}$, there will be exports, while when $P_{int} < P_{aut}$, there will be imports.

Second, for a given difference between the autarky price and the international price, the volume of trade (exports or imports) increases with the size of β or δ. This is intuitive, since a higher β means a flatter demand curve and a higher δ a flatter supply curve. That implies that both consumers and producers respond more to the price change that occurs when we open up for trade, which in turn leads to more exports in the export scenario and more imports in the import scenario.

First, assume that the international price P_{int} is higher than the autarky price P_{aut}, so that we have exports, as illustrated in Fig. 8.12. Producers sell a quantity Q_c such that the marginal cost equals the international price, while domestic consumers buy a quantity Q_b where the marginal willingness to pay, given by the demand curve D, equals the international price. Exports are the difference between production and consumption, $Q_c - Q_b$.

Conrad looks at the figure and is pleased with the opportunities to sell paper abroad. But he is also thinking about the local consumers.

"I see this is good for us producers," says Conrad, "but is it good for society as a whole?" More on that in the next chapter.

Now assume that the international price P_{int} is below the autarky price P_{aut}, so that we have imports, as illustrated in Fig. 8.13. Producers will supply a quantity Q_b such that marginal cost equals the international price, while domestic consumers will buy a quantity Q_c, where the marginal willingness to pay, given by

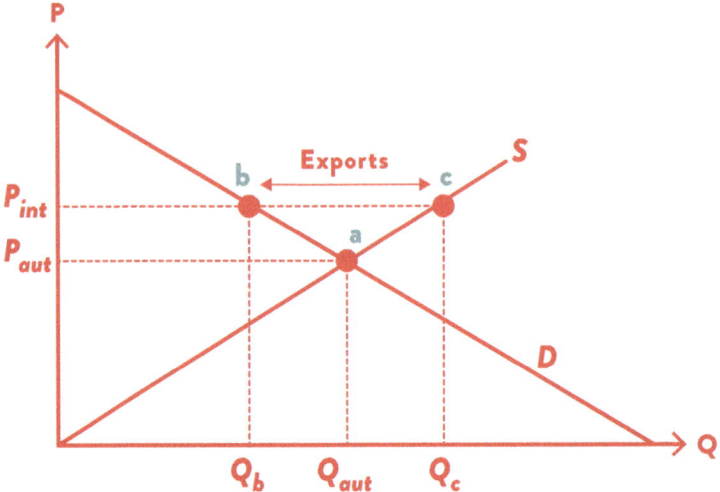

Fig. 8.12 Exports. *Note* The autarky solution is given by point a, with production Q_{aut} and price P_{aut}. When the international price $P_{int} > P_{aut}$, we get exports, where consumers choose point b with consumption Q_b and producers point c with production Q_c, and where the amount of exports is given by the difference between production and consumption

the demand curve D, equals the international price. The import quantity is the difference between consumption and production, $Q_c - Q_b$.

Conrad didn't like the import figure quite as much. Here local producers lose part of the market to foreign producers, and he thinks this certainly can't be good for society. But is he right? You will find the answer in the next chapter.

8.9   Summary

The interaction between supply and demand creates a market equilibrium, a balance point. This gives us a starting point to study shocks in the market, both on the supply and demand sides, and the analysis has given Conrad useful insight into what explains the ups and downs in the price of paper.

We then opened the market to international trade and saw that, depending on how the international price relates to the autarky price, the country will either export or import the good. Conrad was worried that trade would reduce production, but that does not necessarily have to be the case: international demand for paper can provide a basis for increased production in the factory.

The market model is an incredibly useful tool for analysing questions of price formation; it provides answers that would not necessarily be easy to figure out on your own! This is the beauty of a good theory, it creates a structure that helps us think logically about complex problems. The perfect competition model is also

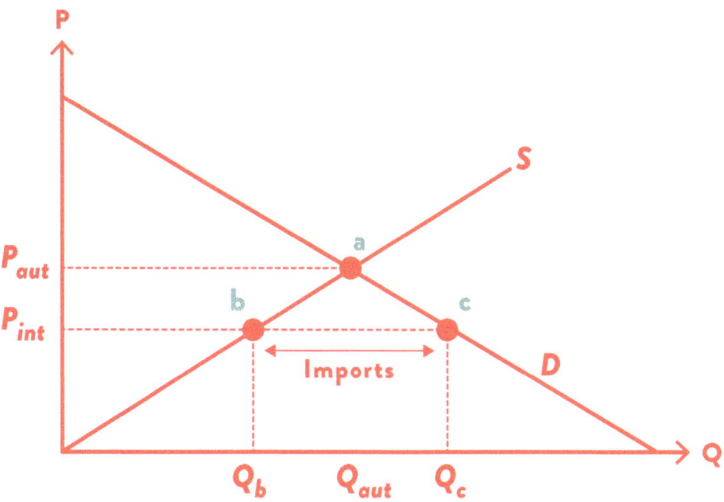

Fig. 8.13 Imports. *Note* The autarky solution is given by point a, with production Q_{aut} and price P_{aut}. With an international price $P_{int} < P_{aut}$, we get imports, where producers choose point b with production Q_b and consumers point c with consumption Q_c, and where the amount of imports is given by the difference between consumption and production

an excellent starting point to discuss what constitutes a good outcome for society, which is the topic for the next chapter—so stay tuned!

8.10   Key Terms

Perfect competition: A market with many producers offering a homogeneous product, where all firms take the market price as given.

Demand elasticity: The percentage change in quantity demanded when the price increases by one percent.

Unitary elastic demand: When a one percent increase in price leads to a one percent decrease in demand.

Elastic demand: When a one percent increase in price leads to a more than one percent decrease in demand.

Inelastic demand: When a one percent increase in price leads to a less than one percent decrease in demand.

Supply elasticity: The percentage change in quantity supplied when the price increases by one percent.

Market equilibrium: The price at which supply equals demand.

Autarky: A situation without international trade.

8.11   Multiple-Choice Exercises

8.1: The Demand Function

Consider the two demand curves in the figure below, based on the demand function $Q^D = \alpha - \beta P$. Assume that the demand is described by D_A for consumer A and D_B for consumer B.

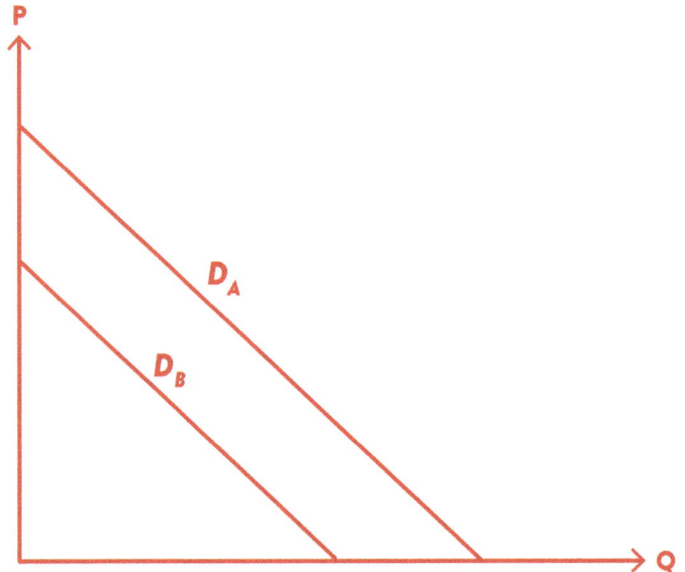

   Compared to consumer B, which of the following statements is true for consumer A?

A. Both α and β are higher for A
B. α is lower, while β is higher for A
C. α is higher, and β is lower for A
D. α is higher for A, while β is the same

8.2: From Individual Demand to Market Demand

Assume A and B are the only consumers in the market for frozen pizza. Consumer A's weekly demand is given by $Q_A = 6 - 0.2P$, and consumer B's is given by $Q_B = 15 - 0.3P$. In the figure below, which curve shows the weekly market demand for frozen pizza?

A. a
B. b
C. c
D. d

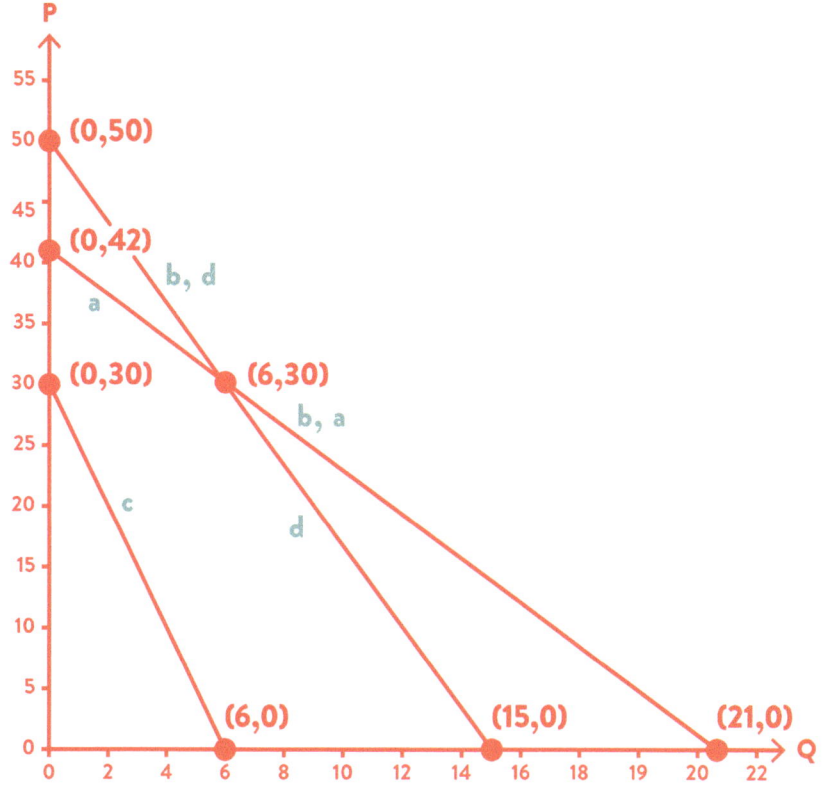

8.3: Shift in Demand and Supply

A pandemic causes a fall in demand for oil, while at the same time, infection control measures make it more expensive to transport oil to markets, which we can interpret as a supply shock. What is the effect of this on the oil price and the quantity of oil sold in the market?

A. The effect on the oil price is uncertain, but the quantity sold goes down
B. Both the oil price and the quantity sold clearly go down
C. Both the oil price and the quantity sold clearly go up
D. The oil price rises, but the effect on the quantity sold is uncertain

8.4: Export and Import

The demand for paper in the home country is given by $Q^D = 75 - 0.5P$, while the supply is given by $Q^S = -30 + 2P$. If the international price $P_{int} = 60$, which of the following statements is true?

A. The home country will import paper
B. The home country will export paper
C. The home country will neither import nor export paper
D. The home country will both import and export paper

Solutions: 8.1 D; 8.2 B; 8.3 A; 8.4 B.

Economic Efficiency

9

Conrad thinks it's wrong that electricity exports lead to higher prices for the country's consumers, but Anna says that this is beneficial for society as a whole.

Conrad sees smoke rising from the chimneys of the paper mill and worries about the environment—and Anna agrees: this is definitely a problem for society.

9.1 Introduction

In the previous chapter, we studied how the interaction between demand and supply creates an equilibrium in a perfectly competitive market, and how this equilibrium is affected by various shocks or international trade.

Conrad likes his factory, but worries about the environment

K. Bjorvatn, *Microeconomics Made Simple*, Classroom Companion: Economics,
https://doi.org/10.1007/978-3-032-06354-0_9

But we said nothing about whether the outcome is actually good for society.

Is it economically sound that the price of electricity rises when there is little rainfall and cold weather sets in? Or that Conrad must reduce his production in response to international competition?

In this chapter, we will get to know *the invisible hand* and introduce one of the oldest and most fundamental insights in economics: that a perfectly competitive market, under certain conditions, provides the best economic outcome for society.

No other price than the market price generates a higher total surplus for consumers and producers combined.

We will then introduce another classic result from economics—comparative advantage and the gains from international trade.

Some might say: *"Imports displace local production, and exports hurt local consumers."*

Yes, international trade is controversial, but we will see that trade creates gains for the country as a whole.

However, the market solution is not always good, and we will examine this in the final part of the chapter. One important source of what economists call *market failure* is negative external effects linked to the environment. Production exceeds the socially optimal level, and this creates economic inefficiency, often referred to as a deadweight loss. In such cases, government intervention through environmental policy is necessary to ensure a good outcome for society—but more on economic policy in Chapter 10.

9.2   The Invisible Hand

When consumers pay less for a good than they are willing to pay, a consumer surplus arises. And when producers receive more than it costs to produce a good, a producer surplus arises. In a market model, the consumer surplus (CS) is the area below the demand curve and above the price, while the producer surplus (PS) is the area above the supply curve and below the price, as shown in Fig. 9.1.

To understand this, recall that the demand curve can be interpreted as the consumers' willingness to pay, and the supply curve represents the marginal cost of production. Where the two curves intersect, the marginal willingness to pay and the marginal cost are identical. But for all units traded up to this point, the willingness to pay is higher than the marginal cost, creating a surplus for both consumers and producers. The sum of these two surpluses is referred to as the economic surplus.

One of the most fundamental insights in economics is that a perfectly competitive market leads to an efficient allocation of society's resources. In Fig. 9.1, the market price P_a results in a quantity Q_a and economic surplus is maximised. We say that perfect competition leads to economic efficiency. This insight is often attributed to the founder of modern economics, Adam Smith (1723–1790). In *The Wealth of Nations* (1776), he writes the following about human motivation and action:

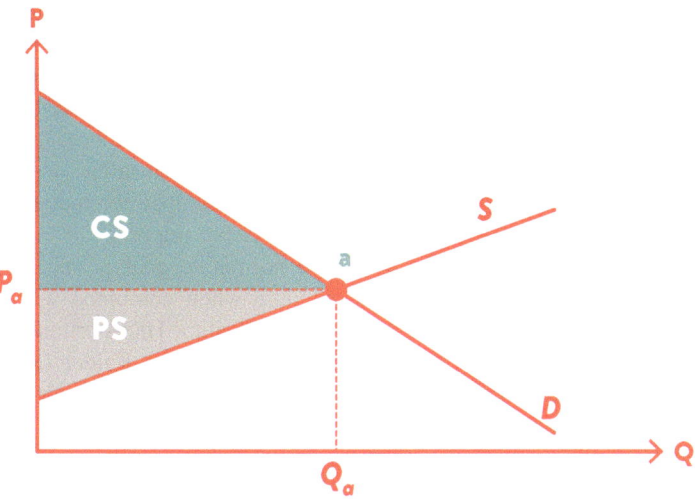

Fig. 9.1 Consumer surplus and producer surplus. *Note* Market equilibrium is given by point *a*, with quantity Q_a and price P_a. The consumer surplus (CS) is the area below the demand curve D and above the price. This area represents a gain for consumers, because up to quantity Q_a, they pay a price that is lower than their willingness to pay, as indicated by the demand curve. The producer surplus (PS) is the area above the supply curve S and below the price. This area represents a gain for producers, because up to quantity Q_a, they receive a price that is higher than their marginal cost, as shown by the supply curve. The sum of the consumer and producer surplus is referred to as economic surplus

> *... he intends only his own gain, and he is in this, as in many other cases, led by an invisible hand to promote an end which was no part of his intention.*

Consumers maximise their utility and producers maximise their profit, and the sum of these choices—motivated by self-interest—results in an economically efficient outcome, as if guided by an invisible hand.

However, the market does not always succeed in creating an efficient outcome on its own. In this book, we will look at two important sources of what economists call market failure.

The first example of market failure is monopoly, that is, a market with only one producer. A monopolist has an incentive to reduce the quantity sold to raise the price and increase profits. That may be good for the producer, but it's bad for consumers, and bad for society. More on monopoly in Chapter 11.

The second example of market failure is external effects, or externalities. The problem here is that producers do not consider the full economic cost of their production. Pollution is a classic example of a negative externality, and we will return to this at the end of the chapter.

Another issue is that government policy may interfere with the work of the invisible hand, thereby creating economic inefficiency. This is an important topic in Chapter 10.

Some scepticism about whether perfect competition is truly desirable stems not from concerns about market failure, but from disagreement over what is meant by "desirable." In this book, we use economic efficiency as the standard for evaluating how well the market performs. But what about the distribution between consumers and producers—or between different consumers and different producers? Is the price that arises in a market fair? A high price may seem unfair to consumers, especially those with low incomes. And international competition may seem unfair to local producers who struggle to compete with cheap imports.

One defence of economic efficiency as a normative goal is that, at least in principle, it provides the best possible outcome when combined with sensible redistribution policies. This is the idea behind the Kaldor-Hicks criterion for economic efficiency, named after British economists Nicholas Kaldor (1908–1986) and John Richard Hicks (1904–1989): Make the pie as big as possible—and then think about how to divide it.

9.3   Comparative Advantage

A fundamental result in economic theory is that international trade is beneficial: it contributes to greater economic efficiency. This insight dates to the British economist David Ricardo (1772–1823).

Ricardo became interested in economic theory after reading Adam Smith's *Wealth of Nations*, and in 1817, he published *The Principles of Political Economy and Taxation*, in which he developed the theory of comparative advantage. He writes:

> *England may be so circumstanced, that to produce the cloth may require the labour of 100 men for one year; and if she attempted to make the wine, it might require the labour of 120 men for the same time. England would therefore find it in her interest to import wine and purchase it by the exportation of cloth.*

In Ricardo's classic example, comparative advantage involves England and Portugal buying and selling cloth and wine. Imagine that consumers in the two countries have the same needs for the two goods—let's say 100 pieces of cloth and 100 bottles of wine each. Assume that England would need 100 labour-years to produce the cloth it needs and 120 labour-years to produce the wine, while Portugal would need only 90 labour-years for the cloth and 80 labour-years for the wine, as shown in Table 9.1.

We see that Portugal is more efficient than England in producing both cloth and wine—so one might wonder whether Portugal really has anything to gain from trading with England!

In Ricardo's example, countries will specialise production according to their comparative advantage. How many hours of labour does each country need to produce cloth relative to what it needs to produce wine? The country with the lowest relative labour requirement in cloth has a comparative advantage in cloth.

Table 9.1 Autarky

	Cloth	Wine	Labour
England	100	120	220
Portugal	90	80	170
Total	190	200	390

Note The table shows the number of work hours needed to produce cloth and wine in the two countries in autarky, that is, when there is no trade between them

Similarly, the country with the highest relative labour requirement in cloth has a comparative advantage in the other good, that is, wine.

From Table 9.1 the relative labour requirement for producing cloth in terms of wine in England is:

$$\frac{Cloth}{Wine} = \frac{100}{120} \approx 0.83 \quad \text{Relative cost of producing cloth in England}$$

While in Portugal it is:

$$\frac{Cloth}{Wine} = \frac{90}{80} = 1.125 \quad \text{Relative cost of producing cloth in Portugal}$$

The cost of producing cloth is therefore relatively lower in England than in Portugal, and we can conclude that England has a comparative advantage in the production of cloth, while Portugal has a comparative advantage in wine. With trade, each country will specialise in the production of the good in which it has a comparative advantage, export that good, and import the other.

With specialisation of production based on comparative advantage, England will produce only cloth, and Portugal only wine. If we allocate all production of cloth to England, it will take 200 labour-years to cover the global (i.e., England and Portugal's) demand for this good—100 pieces of cloth for English consumers and 100 pieces of cloth exported to Portuguese consumers. Similarly, if we allocate all wine production to Portugal, it will take 160 labour-years to meet global demand for wine—100 bottles for consumers at home in Portugal and another 100 bottles exported to England. This is shown in Table 9.2.

Table 9.2 Trade

	Cloth	Wine	Labour
England	200	–	200
Portugal	–	160	160
Total	200	160	360

Note The table shows the number of work hours needed to produce cloth and wine in the two countries when they specialise production according to their comparative advantages

The gains from trade lie in the labour saved. While England, in autarky, used 220 labour-years to produce what it needed, specialising according to its comparative advantage and opening for trade, it now uses only 200. Similarly, Portugal saves 10 labour-years by specialising in wine production.

As a result, the world saves 30 labour-years through trade—which can, for instance, be used for leisure or other enjoyable activities. What you have now learned is indeed one of the deepest insights in economics: trade brings mutual gains! This also carries a powerful message in a world where trade is often seen as a zero-sum game, where one country's gain necessarily implies another country's loss. Ricardo demonstrated more than two centuries ago that this is simply not the case!

9.4   Winners and Losers from International Trade

While international trade creates gains for countries, it also creates winners and losers within countries, which is why trade remains controversial—as we will now see. You remember we discussed international trade in Chapter 8, right? At that point, we carried out thought experiments involving exports and imports, but without saying much about winners and losers.

International trade is controversial

We now analyse consequences of trade for economic efficiency and distribution within that framework. And just like in Sect. 8.8, we divide the discussion into

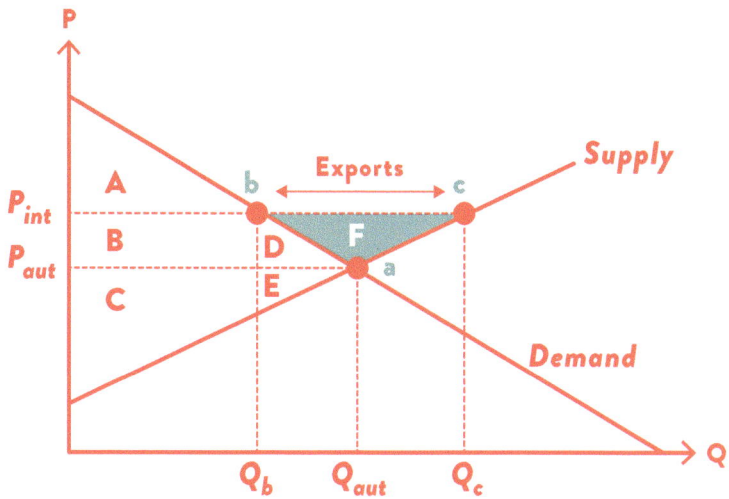

Fig. 9.2 Winners and losers from exports. *Note* This figure builds on Fig. 8.12, which illustrates the exports scenario. The autarky outcome is given by point a, with production Q_{aut}, and price P_{aut}. With an international price $P_{int} > P_{aut}$, the country becomes an exporter, with exports equal to the difference between production Q_c and consumption Q_b. Compared to autarky, exports lead to a gain for producers equal to areas BDF, where BD is transferred from consumers, while F results from export. F thus represents the economic gain from trade in the exports case

two scenarios, depending on whether the country is an exporter or importer of the good.

Let's begin with the exports scenario and look at Fig. 9.2, where in autarky the market settles at point a, with price P_{aut}. Consumer surplus is then the area ABD, and producer surplus is CE.

With the international price P_{int}, consumers choose point b and producers point c.

Consumer surplus is now A, while producer surplus is BCDEF. Trade causes consumers to lose BD, and this loss has been transferred to the producers. This is a distributional effect of international trade: in the case of exports, producers are the winners and consumers the losers.

Compared to autarky, a new area of economic gain appears—namely area F.

This surplus accrues to the producers and results from selling to foreign buyers at a price higher than the marginal cost. And this is the economic gain from trade in the export scenario.

Conrad feels that exporting electricity, which drives up prices for local consumers, can't possibly be good for society. And yes, consumers may long for the days of autarky, when prices were lower—but for society, that would be a bad idea.

Economists will often say (like Kaldor and Hicks): Let the market do its job and open for free trade—and then use taxes and transfers to redistribute the gains.

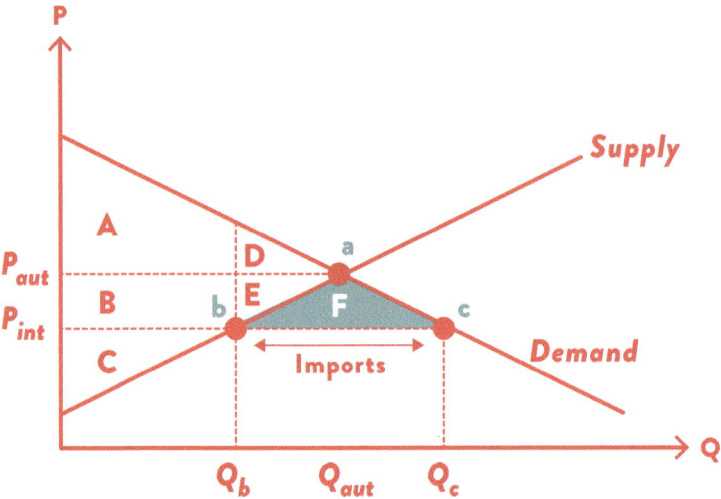

Fig. 9.3 Winners and losers from import. *Note* This figure builds on Fig. 8.13, which illustrates
the imports scenario. The autarky outcome is given by point *a*, with production Q_{aut} and price P_{aut}.
With an international price $P_{int} < P_{aut}$ the country imports the good, with imports equal to the difference between consumption Q_c and production Q_b. Compared to autarky, consumers gain areas
BEF, where BE is taken from producers and F results from import. F thus represents the economic
gain from trade in the imports case

Now assume that the international price is lower than the autarky price, as
illustrated in Fig. 9.3. The autarky equilibrium is again indicated by point *a*, with
consumer surplus equal to AD and producer surplus equal to BCE.

We now see that trade liberalisation leads to imports. Producers choose point *b*
and consumers point *c*. Producers now receive a surplus of C, while the consumer
surplus is ABDEF.

Compared to autarky, producers have lost BE, and this area has been captured
by consumers. This is a pure distributional effect: in this scenario, consumers are
the winners at the expense of producers.

In addition, we see that consumers gain area F, which arises from access to the
good at a low international price. Area F therefore represents the economic gain
from trade in the case of imports.

In this situation, producers may long for autarky, free from uncomfortable foreign competition, enjoying a high price and a large producer surplus. But what is
best for society is not to give in to the autarky enthusiasts.

Note that the greater the difference between the international price and the
autarky price, the larger the economic gain from trade. We can see this by shifting
P_{int} upward in Fig. 9.2 and downward in Fig. 9.3—in both cases, the area *F*, which
represents the economic gain, becomes larger. The least beneficial international
price from an economic perspective is the one that equals the autarky price, $P_{int} -
P_{aut}$, because in that case, there are no gains from trade at all!

At the same time, while the gains from trade increase with the price difference between P_{aut} and P_{int}, it is also true that the distributional effects become larger.

Consumer surplus shrinks as P_{int} rises in the exports scenario, and producer surplus shrinks as P_{int} falls in the imports scenario. There is thus a tension between efficiency and distribution in international trade.

9.5 Externalities

Sometimes paper mills and other factories pollute the environment. In this way, they impose a cost on society—we can think of this as the cost of cleaning rivers from toxins and planting trees to neutralise CO_2 emissions in the atmosphere. This is a cost to society, but not a cost that the firms consider in their accounts. We call this an external effect, or externality.

Pollution is an example of a negative externality. But there are also examples of positive externalities, such as when firms create knowledge that benefits society without being paid for it. Or when a student asks a question during a lecture and where the teacher's answer helps fellow students (who probably were wondering about the same thing) understand better.

Assume that every unit produced creates an environmental cost (e). The firm's marginal cost is S, but the marginal cost for society is then $S + e$. We see in Fig. 9.4 that the market outcome at point a is not an economically efficient solution: the last unit produced costs society much more than consumers are willing to pay for it. This is an example of market failure.

The invisible hand simply fails to create an efficient outcome for society in this case.

If producers had considered the full costs of production, including pollution, they would choose point b, where the social marginal cost equals the marginal willingness to pay. This yields the greatest economic surplus (ES), which is the sum of consumer surplus and producer surplus minus the environmental cost.

The efficiency loss, or deadweight loss, from producers choosing a instead of b is given by the area †. You can clearly see this by starting at the ideal point b, then imagining moving toward the market equilibrium a. Each additional unit produced along this path causes an economic loss: it costs more than it's worth!

The costs to society are given by the $S + e$ curve, and the value (i.e., willingness to pay) is given by the D curve. When we reach point a, the gap between the two curves forms the triangle †. This represents the efficiency loss of the market solution compared to the ideal solution.

Pollution, and external effects in general, are an example of market failure, a situation where the market left to itself does not lead to an economically efficient outcome. Positive externalities can also be relevant, for instance when production causes learning that increases the human capital in society. Multiple-choice question 9.4 at the end of this chapter and Exercise 9.5 in the Workbook allow you to dig deeper into this case.

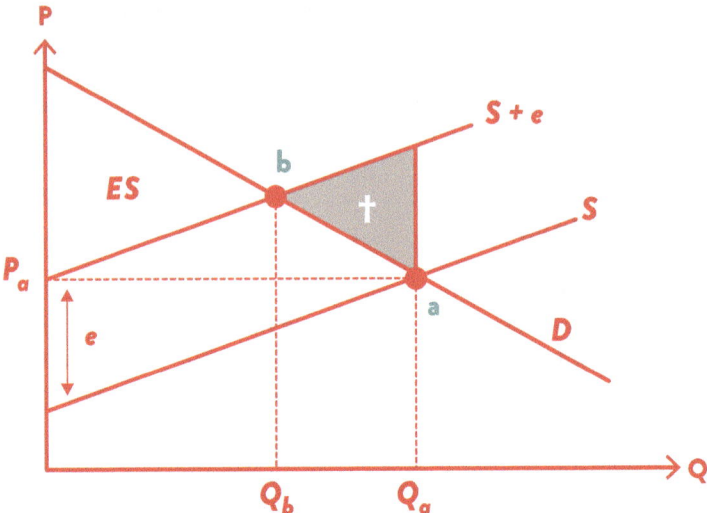

Fig. 9.4 Negative externality and deadweight loss. *Note* When production causes pollution, a discrepancy arises between the marginal cost that producers consider, given by the supply curve S, and the marginal cost to society, given by $S + e$, where e is the environmental cost per unit produced. The market solution is at point a, with production Q_a, while the economically efficient solution is where the social marginal cost $S + e$ intersects demand, at point b, which gives quantity Q_b. The deadweight loss is represented by the shaded triangle † and the economic surplus in the optimal solution is given by ES.

Market power in the form of monopoly and oligopoly are other sources of market failure. More on this in the final part of the book.

9.6   Summary

Whether the market results in a high or low price, the market solution (under certain conditions) is economically efficient, in the sense that consumer surplus and producer surplus are maximised. This is a classic insight in economics, dating back to Adam Smith, who in the 1700s wrote about the invisible hand.

We have also seen that international trade creates gains for the country. This is another classic insight, attributed to David Ricardo, who took over the baton from Adam Smith as the leading economist of his time, in the early 1800s.

At the same time, international trade creates distributional challenges. A free trade policy that leads to exports makes producers winners while consumers lose. To compensate consumers, authorities can impose a tax on producers' surplus and transfer it to consumers, for example through public investments in health and education.

Similarly, producers can be compensated for a free trade that results in imports by lower business taxes or investments in infrastructure that increase business profitability.

In this way, all groups in society can share in the gains from trade, which is important for securing support for a free trade policy.

Finally, we studied environmental issues as an example of market failure, a situation where the market left to itself does not provide an economically efficient solution.

A deadweight loss arises because producers do not take into account the cost their polluting production imposes on society. The invisible hand needs a helping hand—we need environmental policy. And there will be more on this in the next chapter.

9.7   Key Terms

Consumer surplus: The difference between consumers' willingness to pay and what they actually have to pay for a good.

Producer surplus: The difference between the cost of producing a good and what producers actually receive for it.

Economic surplus and economic efficiency: Economic surplus is the sum of consumer surplus and producer surplus. Economic efficiency is achieved when the economic surplus is as large as possible.

Market failure: When the market, left to itself, does not generate the economically most efficient outcome.

Externality: Consequences (costs or benefits) of actions that economic agents do not take into account when making their decisions. A negative externality imposes a cost on society, for instance caused by pollution, while a positive externality generates benefits, one example being innovations from research and development that can be copied by others.

Efficiency loss, deadweight loss: The reduction in economic surplus compared to the economically most efficient outcome.

9.8   Multiple-Choice Exercises

9.1: Consumer Surplus

Assume consumer A's demand for frozen pizza is $Q_A = 6 - 0.2P$ and consumer B's demand is $Q_B = 15 - 0.3P$.

What is the consumer surplus in the market if the market price is 30 euros?

A. 20
B. 40
C. 60
D. 80

9.2: Economic Efficiency and Distribution

Look at the figure below. Imagine that the authorities initially set a price ceiling P_b on the good so that the market outcome is at point b.

What are the consequences of removing this price ceiling so that equilibrium moves to point a?

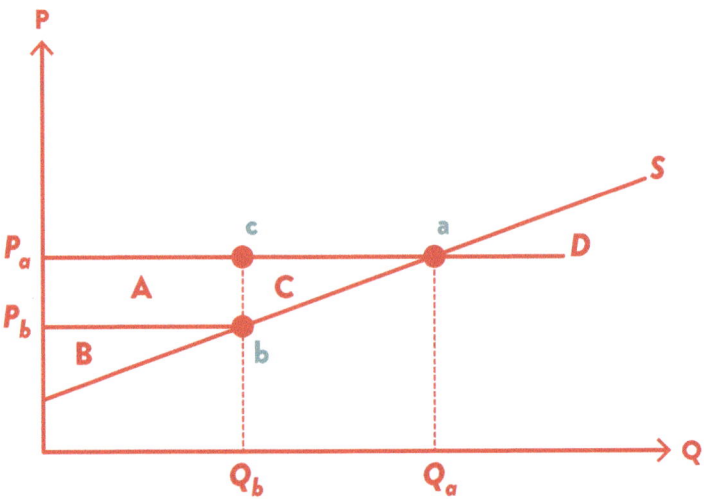

A. Society gains C, consumers lose A
B. Society gains ABC
C. Producers gain ABC
D. Society gains C, producers gain A

9.3: International Trade and Economic Efficiency

A reduction in the international price of a good will:

A. Always result in an economic gain
B. Never result in an economic gain
C. Result in an economic gain only if the country imports the good
D. Result in an economic gain only if the country exports the good

9.4: Deadweight Loss from a Positive Externality

Consider the figure below. The supply curve S and demand curve D intersect at point a, which is the market equilibrium. But production creates a positive external effect for society's human capital, with value h for each unit produced. An example could be that firms contribute knowledge through their production that benefits the entire society, so the social marginal cost is $S - h$.

How large is the deadweight loss (†) associated with the market equilibrium in this case?

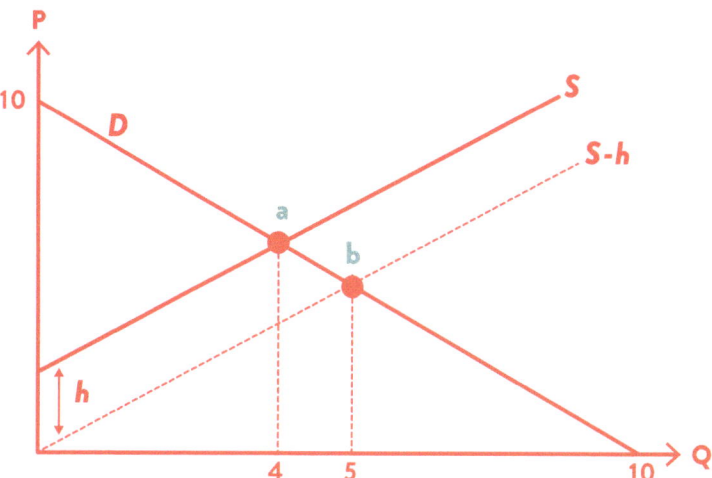

A. $† = 16$
B. $† = 5$
C. $† = 1$
D. $† = 2$

Solutions: 9.1 C; 9.2 A; 9.3 C; 9.4 C

Economic Policy

10

Conrad is worried that a new tax on paper will hurt the factory's profitability. Anna tries to reassure him: much of the burden will fall on consumers.

Brian claims that a lower tax rate can lead to higher tax revenue. The Tax Enthusiasts are scratching their heads.

10.1 Introduction

Conrad is worried. He has heard that the government is considering introducing a tax on paper. But who will end up paying it? Conrad and the other producers—or will they be able to pass the tax on to consumers?

Conrad reads in the paper about a new tax on paper

© The Author(s), under exclusive license to Springer Nature Switzerland AG 2026
K. Bjorvatn, *Microeconomics Made Simple*, Classroom Companion: Economics,
https://doi.org/10.1007/978-3-032-06354-0_10

This chapter is mainly about taxation. Taxes are necessary to fund healthcare, education, and national defence, as well as to support those in need—such as families with young children, the unemployed, and pensioners. But tax policy is also highly contested. The political left and right disagree on how high taxes should be, what should be taxed, and how tax revenues should be used. One reason why taxation is so controversial is that it affects both the distribution of economic surplus and economic efficiency.

Who bears the brunt of the tax—consumers or producers? Different political parties may prioritise the interests of one group over the other. And when a tax distorts an otherwise efficient market, it creates an economic cost—a deadweight loss—and people may disagree on how serious this is compared to other policy goals.

Whatever your political views, it's important to understand the effects of economic policy on both efficiency and equity—and that's what this chapter is all about.

Some types of taxes, however, should be relatively uncontroversial. Think of pollution and the environment, as discussed in the previous chapter. Here, the market has a problem: a negative externality leads to deadweight loss. Against this backdrop, the environmental tax appears as the ideal remedy. It improves environmental outcomes, raises revenue for the government, and promotes economic efficiency—a win-win-win situation. Whether your political leanings are green, red, or blue, this kind of tax is simply good policy.

Although we shall focus on taxation in this chapter, we will also look at other types of government intervention— such as subsidies, trade policy, quotas, and price regulations. Subsidies can be used to encourage environmentally friendly behaviour. Tariffs are often proposed to protect domestic industries. Price regulations appear in the labour market in the form of minimum wages, and as rent control in urban areas. Quantity regulations, such as quotas and licensing schemes, are common in markets where the authorities want to limit entry—whether for environmental reasons or to protect producers.

We'll cover all of this in the current chapter. But we begin with Conrad's concern about the new paper tax: How badly will it affect the factory's profits? And what will the tax mean for overall economic efficiency?

10.2   Taxes and Tax Incidence

Conrad reads in the newspaper that the government has proposed a new tax on paper. It will be levied on producers, and Conrad believes this is deeply unfair. He worries that it will seriously hurt the profitability of the paper mill.

"New tax on paper." Conrad is concerned that the new tax will hurt the factory's profits

Anna hopes a dose of economic theory will calm him down. "It won't be that bad, Grandpa. Consumers will end up paying for at least part of it."

Still, Conrad insists that taxing producers is anti-business, and declares he'll never vote for the governing party again. But Anna replies that it doesn't matter whether the tax is levied on producers or consumers—the outcome is identical!

She asks Conrad to sit down, grabs a pen and a piece of paper, and starts drawing Fig. 10.1.

A production tax feels to producers like an increase in costs. They not only have to pay for labour and raw materials, for each unit they produce, they must also pay a tax to the government.

The marginal cost including tax is now $S + t$, and we can understand Conrad's concern. While the last unit produced at the original market equilibrium a was just marginally profitable (since the marginal cost equalled consumers' marginal willingness to pay), with the tax, that final unit becomes clearly unprofitable.

Conrad and his colleagues therefore must cut back on paper production to reduce costs.

But by how much?

Conrad thinks production must be cut all the way down to point d, since this is where the marginal cost including the tax equals P_a, the market price before the paper tax. But Anna explains that it won't be that bad. As output falls and we move down along the marginal cost curve, we're also climbing up the demand curve.

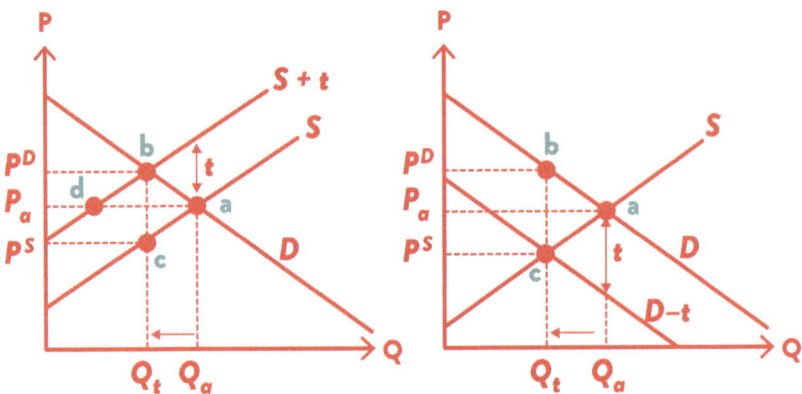

Fig. 10.1 **a** A tax on producers. **b** A tax on consumers. *Note* Fig. 10.1a shows the effect of a tax on producers, where the pre-tax equilibrium is at point *a*. The tax *t* represents an extra cost that shifts the supply curve upwards to $S + t$. The new equilibrium is found where the marginal cost including the tax intersects the demand curve—at point *b*, with quantity Q_t. The price paid by consumers is P^D, while the price received by producers is P^S where $t = P^D - P^S$. Fig. 10.1b illustrates an equivalent tax *t* levied on consumers. After the tax, consumers' willingness to pay is effectively reduced, and this can be represented by the demand curve shifting down to $D - t$. The new equilibrium is at point *c*, with quantity Q_t. The consumer price and producer price are the same as in the case of a producer tax, so it makes no difference who formally pays the tax—the economic outcome is identical

This is the effect Conrad had overlooked: a lower quantity means a higher willingness to pay, so producers are able to pass on some of the tax burden to consumers. The new equilibrium is at point *b*, where $S + t$ intersects the demand curve *D*, and the quantity bought and sold has been reduced to Q_t. Consumers pay a price P^D, and producers receive a price P^S, where $t = P^D - P^S$.

How much the willingness to pay increases as quantity falls depends on the slope of the demand curve. In Fig. 10.1a, we see that the tax causes the consumer price to rise by about as much as the producer price falls, so the tax burden is roughly shared equally between consumers and producers. But that's not always the case—it depends on the slopes of the supply and demand curves: the one with the steeper curve pays most of the tax.

Let's now look at Fig. 10.1b, which illustrates a very interesting insight from market theory: it doesn't matter whether the tax is levied on producers or consumers—the outcome is the same!

A tax on consumers can be interpreted as a leftward shift of the demand curve, from *D* to $D - t$. This represents consumers' willingness to pay after tax. Since the last unit consumed at point *a* only gave a marginal gain in consumer surplus, consumers are clearly not interested in buying the same quantity when a tax is added to the price.

Those with the lowest willingness to pay therefore reduce their consumption, and just like in the discussion of a producer tax, two things happen: we move up along the demand curve and down along the supply curve. This process only stops at quantity Q_t, where consumers pay a price P^D and producers receive a price P^S—exactly as in the case of a producer tax. We see that no matter who initially pays the tax, the buyer and the seller end up sharing the burden!

Since it doesn't matter whether we shift the supply curve or the demand curve in the analysis, we can instead represent the tax as a wedge, as shown in Fig. 10.2. Compare this with what you saw in Figs. 10.1a and b—the consequences of the tax are the same. And since the figures are easier to read without all the shifting curves, we'll stick to the wedge method for the rest of the chapter.

While in a perfectly competitive market it doesn't matter whether the tax is imposed on producers or consumers, this doesn't necessarily mean that consumers and producers bear an equal share of the tax burden. A more detailed analysis is found in Math Box 10.1, and as you will see, tax incidence—that is, who ends up paying the tax—depends on the slopes of the supply and demand curves.

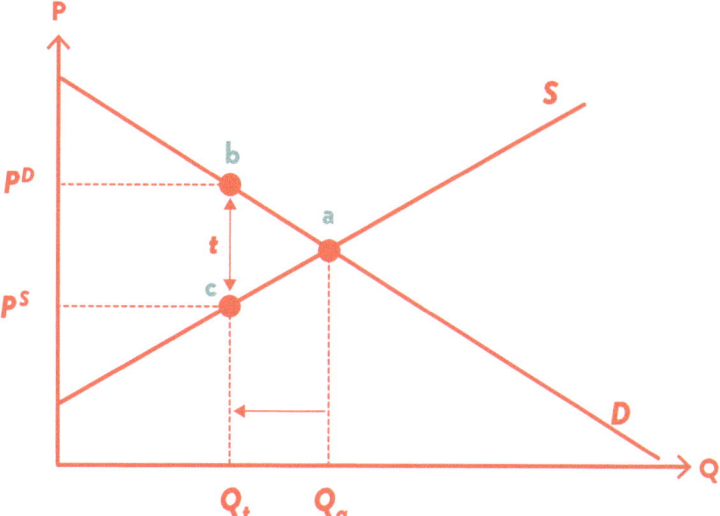

Fig. 10.2 The tax wedge. *Note* Fig. 10.2 shows how a tax creates a wedge between demand and supply, and thus a gap between the consumer price P^D and producer price P^S

Math Box 10.1: Tax Incidence

A per-unit tax of size t creates a wedge between the price consumers pay (P^D) and the price producers receive (P^S):

$$P^D = P^S + t$$

In equilibrium, the quantity demanded equals the quantity supplied $Q^D = Q^S$. With the demand function $Q^D = \alpha - \beta P^D$ and the supply function $Q^S = -\gamma + \delta P^S$ we get the following equilibrium condition:

$$\alpha - \beta P^D = -\gamma + \delta P^S$$

We know that $P^D = P^S + t$ and plug this into the expression above:

$$\alpha - \beta(P^S + t) = -\gamma + \delta P^S$$

The producer price in equilibrium is therefore:

$$P^S = \frac{\alpha + \gamma}{\beta + \delta} - \frac{\beta t}{\beta + \delta}$$

The first term on the right-hand side is the equilibrium price *without* the tax, P^*, (see Math Box 8.1), so the producer price with the tax becomes:

$$P^S = P^* - \frac{\beta t}{\beta + \delta} \quad \text{Producer price with tax}$$

Substituting this into $P^D = P^S + t$ we find the consumer price:

$$P^D = P^* + \frac{\delta t}{\beta + \delta} \quad \text{Consumer price with tax}$$

The second term in the expressions above shows how a tax affects the producer price and the consumer price—this is what we refer to as the tax incidence: who bears the burden of the tax?

We see that a higher β, which corresponds to a more price-sensitive (flatter) demand curve, means the tax has a greater negative impact on the producer price. Conversely, a higher δ, which corresponds to a more price-sensitive (flatter) supply curve, means the tax leads to a larger increase in the consumer price.

Intuitively, with a flat demand curve, it's difficult to get consumers to pay more—so producers end up bearing most of the tax burden. Similarly, with a flat supply curve, producers can't lower their prices by much, and consumers must therefore shoulder most of the cost.

An interesting application of the theory of tax incidence is the electricity market, which features a step-shaped supply curve—in Fig. 10.3 shown with just one step. Think about an energy system based entirely on hydro power, with zero marginal cost up to a capacity limit, Q_0, beyond which the supply becomes vertical. Figure 10.3a shows the market in summer when demand is low since the weather is warm, while Fig. 10.3b shows the market in winter, when demand is high, as people need to heat up their houses.

Let's begin with a situation without a tax. In summer, the market equilibrium occurs at point a with price $P_a = 0$, while in winter the price is higher, $P_a > 0$. Note that the summer equilibrium lies on the horizontal part of the supply curve (there is plenty of water in the reservoirs), while in winter, we are on the vertical part (all the water has been used up).

What happens if we introduce a tax—say, an electricity tax, t? We insert a tax wedge between the horizontal segment of the supply curve and the demand curve and find a new equilibrium at point b, where the consumer price has increased to P^D, while the producer price is P^S, that is, the same as before the tax, P_a. Referring to Math Box 10.1, a horizontal supply curve means $\delta = \infty$, which implies consumers bear the entire burden when a tax is introduced.

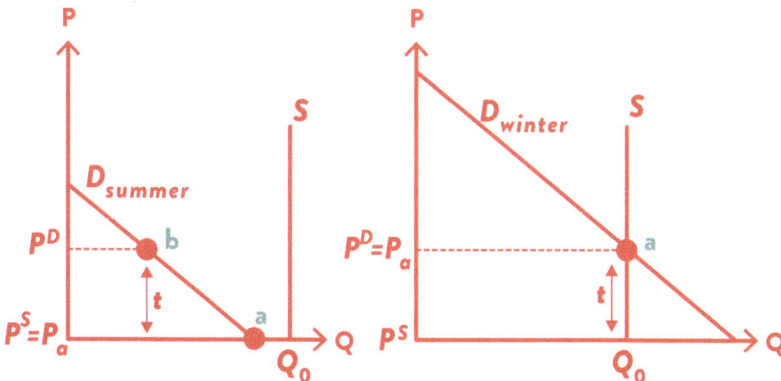

Fig. 10.3 Electricity tax in summer and winter. *Note* Fig. 10.3a shows the market in summer, where warm weather leads to low electricity demand and equilibrium at point *a* with a price of zero. An electricity tax in summer shifts the equilibrium to point *b*, where the consumer price has increased while the producer price remains unaffected. With a horizontal supply curve, consumers bear the entire tax burden. Figure 10.3b shows the market in winter, where cold weather causes higher demand and a higher price, $P_a > 0$. An electricity tax in winter, when the supply curve is vertical, does not affect the consumer price at all, and producers bear the full tax burden

In winter, the picture is different. A vertical supply curve means the quantity supplied is unresponsive to price, given by $\delta = 0$ in Math Box 10.1, and producers bear the entire tax burden. As shown in Fig. 10.3b, the tax wedge brings the producer price down to zero while the consumer price remains unchanged, that is, $P^D = P_a$. The electricity tax thus has no effect on the consumer price in winter.

Although supply curves are rarely perfectly horizontal or vertical, this example highlights an important point: tax incidence depends on the steepness of the supply curve relative to the demand curve, with the side having the steepest curve bearing most of the tax. Intuitively, a steep curve implies limited flexibility, which is clearly a disadvantage if you want to avoid paying a tax.

10.3   Taxes and Economic Efficiency

The Efficiency Party cares more about efficiency than distribution. They understand that the government needs revenue but want to minimise the damage taxes cause to the economy. Taxes typically have side effects—they create deadweight loss. In fact, subsidies do the same, as we will soon see. Figure 10.4a shows the deadweight loss caused by a tax. The starting point is market equilibrium at point a, with price P_a and quantity Q_a. Enter now a tax, t, creating a wedge between the consumer price P^D and producer price P^S and a reduction in the quantity bought and sold to Q_t. This is the analysis we recognise from Fig. 10.2. And since the supply curve is quite steep, producers must pay most of the tax.

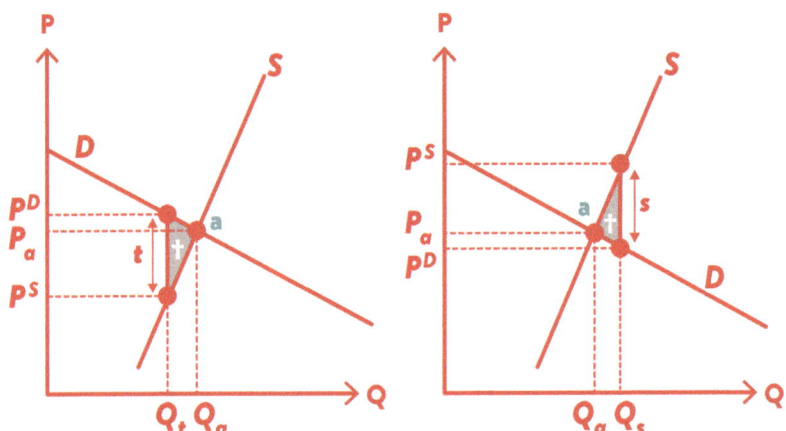

Fig. 10.4 a Tax. **b** Subsidy. *Note* Fig. 10.4a shows how a tax t affects the market and creates a deadweight loss (†), while Fig. 10.4b shows the corresponding effect of a subsidy s. Because the supply curve is steep (and steeper than the demand curve), producers bear most of the tax burden (we see that the producer price, P^S, has fallen significantly), but producers also gain the most from a subsidy (producer price P^S rises substantially)

What's new in Fig. 10.4a is that it also shows the deadweight loss indicated with †. The tax displaces purchases and sales that would have created surplus for consumers and producers, causing a loss of economic surplus. From the figure, you see the loss triangle is quite small, because when supply is inelastic (i.e., not very price-sensitive), the tax has little impact on the quantity bought and sold (only a modest decrease from Q_a to Q_t). The same logic applies if the demand curve is steep: consumers bear most of the tax burden, but the deadweight loss is quite small.

Figure 10.4b shows the case with a subsidy, s, and there are several interesting points when comparing this to the tax scenario. First, a subsidy increases quantity, moving from Q_a before the subsidy to Q_s after. Second, the subsidy makes the producer price higher than the consumer price: $P^S - P^D = s$. Third, while a steep supply curve means producers pay most of the tax, the same steep supply curve means producers receive most of the subsidy. Finally, just as with a tax, a subsidy creates a deadweight loss, but here it arises because too many transactions occur: every unit bought and sold beyond Q_a costs more than it's worth (the supply curve S, showing costs, lies above the demand curve D, showing willingness to pay). The deadweight loss is again small, as the subsidy only slightly changes the equilibrium quantity.

In the figure, the supply curve is the steepest, but if instead the demand curve were steepest, consumers would gain most from a subsidy. We see that with a tax, being inflexible is a disadvantage, while with a subsidy, it is an advantage!

The Efficiency Party likes markets with inflexible consumers and producers: in these markets, taxes can be introduced without causing large efficiency losses. Most economists care about economic efficiency and argue that taxing property is an excellent idea as demand for housing is inelastic (and also from a fairness perspective, since rich people have bigger houses and therefore would pay a higher tax). But here they face strong opposition from the Consumer Party, which advocates for consumer interests, making a tax on housing a politically sensitive topic.

10.4 The Laffer Curve

The relatively small party, the Tax Enthusiasts, advocates higher taxes to strengthen public health care and education, and to raise transfers to the sick and elderly. Most members of the party support a significant increase in taxes without much concern for either efficiency or distribution. Brian (who sympathises with the party but also knows basic economics from his year at the School of Economics) argues that higher taxes do not necessarily lead to more money in the treasury. In fact, he claims that it can make sense to lower taxes to increase tax revenues.

Figure 10.5a shows how a tax t (per unit bought or sold) generates tax revenue τ. In this example, both the lower tax rate t_1 and the higher tax rate t_2 result in the same tax revenue, illustrated by the areas of the two rectangles: $\tau_1 = t_1 Q_1$ and $\tau_2 = t_2 Q_2$. The point is that although increasing the tax rate from t_1 to t_2 raises

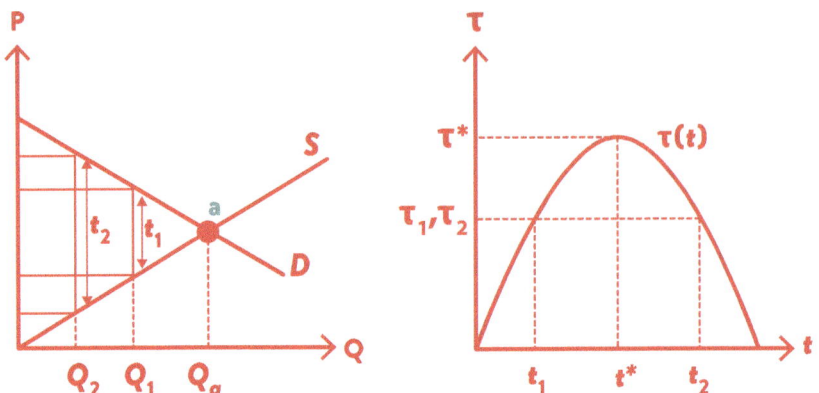

Fig. 10.5 a Tax and tax revenue. **b** The Laffer curve. *Note* Fig. 10.5a shows two different tax rates, a low t_1 and a high t_2, which result in the same tax revenue, illustrated by the areas of the two rectangles. Figure 10.5b shows the relationship between the tax rate t and total tax revenue, $\tau = tQ$. This relationship is hump-shaped and reaches a peak at the tax rate t^*. Depending on where we start, increasing or reducing the tax rate may either increase or decrease total tax revenue

the tax collected per unit, part of the tax base disappears, as indicated by the drop in quantity from Q_1 to Q_2. It is therefore not obvious that a higher tax rate always leads to higher tax revenues.

Figure 10.5b shows tax revenue as a function of the tax rate, where t_1 and t_2 generate the same tax revenue ($\tau_1 = \tau_2$), but where there is a tax level in between, t^*, that yields the highest tax revenue, τ^*.

This hump-shaped relationship is known as the Laffer curve, named after the American economist Arthur Laffer, who is said to have sketched it on a napkin during a meeting with leaders of the Republican Party.

"It's obvious that if taxes become high enough, consumers will stop buying the good—it's practically speaking like a ban," says Brian. "And once consumption disappears, so does the tax base." Taxes must not be set so high that they drive away all the consumers. At the same time, we need taxes to finance public goods, and with zero taxes, there will obviously be zero tax revenue. So, there must be some optimal level of taxation for those of us who want to maximise revenue. "The point is, we mustn't go too far!" he emphasises, giving the most enthusiastic tax advocates something to think about.

Ronald Reagan was inspired by the Laffer curve when launching his tax reforms in the 1980s, arguing that cutting taxes would increase government revenue. Note, however, that this argument relies on the assumption that tax rates were initially above t^* in Fig. 10.5b, like t_2 (which they likely were not, as the tax cuts later led to large deficits in the US federal budget). But in theory, at least, Laffer, Brian, and Reagan have a point.

Math Box 10.2: The Laffer Curve

Let tax revenue be given by $\tau = tQ$. Assume that $Q^D = 1 - P^D$ and that the supply function is $Q^S = P^S$. This simplifies the expressions from Math Box 10.1 by setting $\alpha = \beta = 1$ and $\gamma = 0$. From Math Box 10.1, we know the producer price can be written as:

$$P^S = 0.5 - 0.5t$$

Since the supply curve is $Q^S = P^S$, we have:

$$Q^S = 0.5 - 0.5t$$

The tax revenues (τ) are therefore:

$$\tau = tQ = 0.5t - 0.5t^2$$

This is a hump-shaped function, illustrated in Fig. 10.5b.

To find the tax rate t that maximises total tax revenue, we take the derivative of the revenue function and set it equal to zero:

$$\frac{\partial \tau}{\partial t} = 0.5 - t = 0 \Rightarrow t^* = 0.5$$

Note that both a lower tax than t^* and a higher tax than t^* will lead to *lower* total tax revenue.

So if the tax rate is initially very high, reducing the tax rate could theoretically *increase* total revenue.

With the tax rate t^* the quantity produced is:

$$Q^S = 0.5 - 0.5t^* = 0.5 - 0.5(0.5) = 0.25$$

And the maximum total tax revenue is:

$$\tau^* = t^*Q = 0.5(0.25) = 0.125$$

10.5 Tariffs

We live in troubled times, and the new Protectionist Party has gained momentum. They want to introduce a tariff, that is, a tax on imported goods, both to protect local producers and to raise revenue from abroad. However, they face opposition from the Consumer Party and the Efficiency Party, who argue that a tariff will raise prices and harm consumers and economic efficiency.

Look at Fig. 10.6. It shows a situation where the country faces a low international price P_{int} and is therefore an importer. As you know from our earlier analysis of imports (see Figs. 8.13 and 9.3), consumers are at point c and producers at point b.

A tariff causes the international price to rise from P_{int} to $P_{int} + t$. Consumers now settle at point e, and producers at point d. Do you think the Protectionist Party will be satisfied with this policy? Can you understand the concerns of the Consumer Party and the Efficiency Party?

We note that the tariff causes an increase in the domestic price, which benefits producers and harms consumers, but also creates revenues for the government. For the economy, there is a welfare loss indicated by areas $†_1$ and $†_2$. The former is the loss caused by excessive production, that is, production at a cost higher than the world market price P_{int}, while the latter is caused by the tariff crowding out consumption.

Some of the more enthusiastic protectionists propose a very high tariff that would effectively eliminate all imports. They argue that this would give a strong

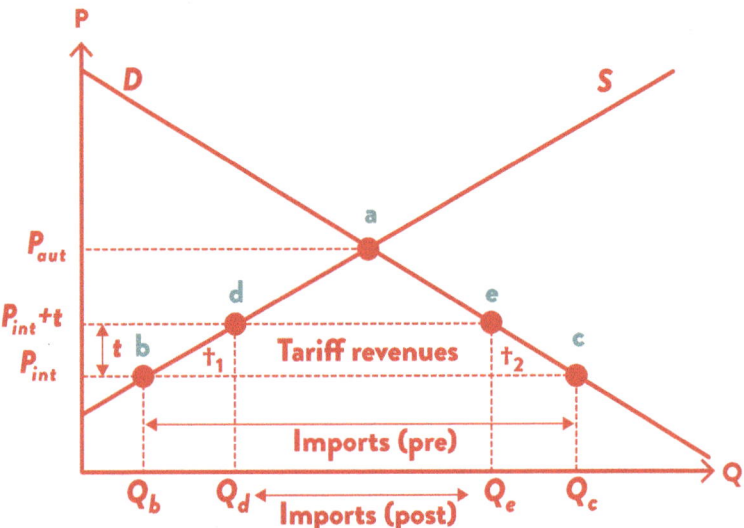

Fig. 10.6 Tariff. *Note* The figure shows a country that initially faces an international price P_{int}, with consumers choosing point c and producers point b, and where imports are the distance between consumption and production. Pre-tariff imports are $Q_c - Q_b$. A tariff t is imposed on imports, raising the import price from P_{int} to $P_{int} + t$. The new consumer equilibrium is now at e, while producers settle at d. The tariff increases production, reduces consumption, and hence also reduces imports, which now, post tariffs, equals $Q_e - Q_d$. Area $†_1$ shows the efficiency loss caused by the increase in production, while $†_2$ shows the efficiency loss caused by the increase in consumption, relative to the socially optimal levels

boost to local producers and generate high tariff revenues. Do you see any flaws in this argument?

10.6   Environmental Policy

As we saw in Chapter 9, the market outcome is not always economically efficient. One important example of market failure is external effects such as pollution. An environmental tax is a good way to address this problem: it improves the environment, increases economic efficiency, and brings revenue to the public budget. A triple win, in other words.

Anna votes for the Green Party but is an economist by training and disposition, and she understands that economic policy is about balancing competing concerns. She draws Fig. 10.7, based on Fig. 9.4. The problem is the deadweight loss † caused by the environmental cost e. Output Q_a is too high and should be reduced to Q_t, where the marginal willingness to pay, as given by the demand curve, equals the social marginal cost $S + e$.

What happens if we now introduce a tax t exactly equal to the environmental cost e? As we know from the discussion on taxation (see Fig. 10.2), this creates a wedge between the consumer price and the producer price. Whereas the market equilibrium before the tax was at point a, with the tax we get a consumer price

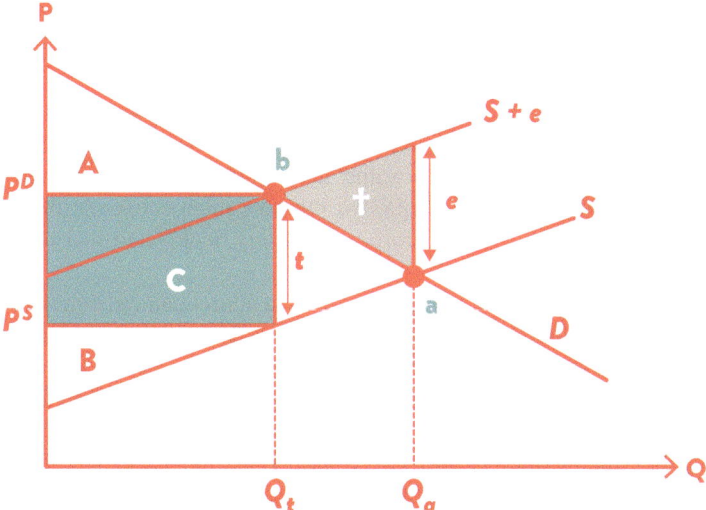

Fig. 10.7   Environmental Externality and Pigouvian Tax. *Note* This figure follows up on Fig. 9.4: each unit produced creates an environmental cost e, and the deadweight loss is given by the area †. A production tax equal to the environmental cost, $t = e$, creates a tax wedge such that consumers choose point b and pay price P^D, while producers receive P^S, and the produced quantity is Q_t. Consumer surplus is area A, producer surplus is area B, and tax revenue is area C. This policy eliminates the deadweight loss and thus leads to a gain in economic efficiency

P^D and a producer price P^S. What will production be under this tax? The answer is: Q_t, and just like that, the deadweight loss is gone!

Let us examine the effect of the tax a bit more closely. We see that the consumer surplus after the environmental tax is A, the producer surplus is B, and the revenue from the environmental tax is t times the quantity Q_t, that is, area C. The environmental tax therefore leads to a redistribution from consumers and producers to the state. This redistribution does not affect economic efficiency. But we also see that the environmental tax has eliminated the deadweight loss †. This is because the tax internalises the external effect: producers must now consider the environmental cost they impose on society.

Even though consumers and producers lose from the environmental tax when viewed in isolation, society gains. The tax thus *removes* rather than creates a deadweight loss. In this case, there is no trade-off between taxation and economic efficiency—the tax leads to an increase in total economic surplus.

We call this type of environmental tax a Pigouvian tax, named after the British economist Arthur Cecil Pigou (1877–1959). Note that pollution is not an argument in favour of shutting down production altogether, as some of the Green Party's most radical activists propose! The Pigouvian tax is an optimal tax in the sense that it balances the concern for the environment with the concern for consumers and producers.

In our earlier example, it was the producers who polluted, and many would argue that in such cases, it is only fair that the producers should pay the tax—not the consumers. This is often referred to as the "polluter pays" principle. But what have we learned from market theory? It doesn't matter whether the tax is imposed on producers or consumers—the outcome is the same! So even if it sounds fairer for producers to pick up the bill, we who have studied microeconomics know that one could just as well impose a tax on consumers to address the problem of smoke coming out of factory chimneys and other negative externalities. The environmental benefits would be the same, and so would the prices faced by both consumers and producers. Once again, we see how microeconomic theory offers insights I guess you didn't have before reading this book—and that most people certainly don't have.

So far, we have considered taxation as a tool to address an environmental issue. The example was pollution, a negative externality. But positive externalities exist as well, and in these cases, the problem is that output is too low.

One example might be when producers generate new ideas that benefit society. If innovation is linked to production—think "learning by doing"—there may be a case for policy intervention to promote overall economic efficiency. In a market with international trade, this can quickly lead to calls for protecting domestic producers by limiting imports. But is protectionism good economic policy to stimulate innovation? You can explore this further in Exercise 10.4 in the Workbook.

10.7   Quantity and Price Regulation

We have spent quite some time discussing taxation as an example of economic policy—and for good reason. Governments need revenue to finance publicly provided goods and transfers and therefore impose taxes on goods and services. But there are also other motives for intervening in the market, for example to address environmental issues, as discussed above. And sometimes the authorities choose to act more directly, using price and quantity regulation rather than taxes and subsidies.

Let's start with quantity regulation and stay with the environment. We've seen that an environmental tax can be an effective policy against pollution. Now I want to show you that (tradable) permits work in the same way as a tax.

Look at Fig. 10.8. Assume that producing one unit results in one unit of CO_2 emissions, and that each CO_2 permit allows a firm to emit one unit of CO_2. By issuing Q_{low} permits to producers, the authorities make sure that production will not exceed Q_{low}. With production limited to Q_{low}, firms can sell their product for P_{high}.

Competition for the permits ensures that only the lowest-cost producers can afford to pay the permit price and still operate profitably. With production limited to Q_{low}, the permit price k equals the difference between the consumers' willingness to pay and the marginal cost. For the last unit produced, the cost—including the permit price—equals the consumer's willingness to pay. The resulting consumer surplus is area A, producer surplus is area B. We can see that the permit price k works exactly like a tax. The government revenue, area C in the figure, is the same as it would have received by imposing a tax.

The assumption that the permits are tradable is crucial. Imagine the government instead randomly distributed the permits and did not allow trading. Then there is

Fig. 10.8 Emissions permits. *Note* In the original market equilibrium, the outcome is at *a*, with quantity Q_a. Introducing tradable CO_2 permits limits production to Q_{low}, with a permit price of k. The consumer price is now P_{high}, while the producer receives P_{low}. Consumer surplus is given by area A, producer surplus by area B, and the government's revenue from selling the permits is area C

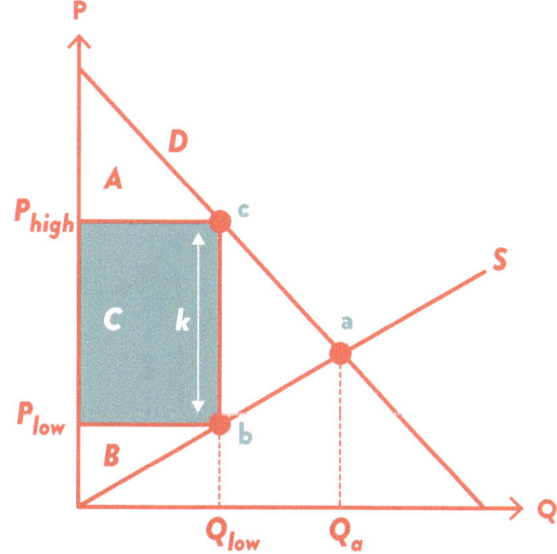

no guarantee that only low-cost producers would be active in the market. This leads to a loss in overall economic efficiency. The multiple-choice question 10.4 at the end of the chapter and Exercise 10.5 in the Workbook provide examples of this: What if fishing quotas were distributed by lottery to prevent overfishing, or traffic into the city centre were regulated using vehicle registration numbers, so-called odd–even driving?

Price regulation is frequent in the labour market, where many countries have imposed minimum wages. I encourage you to look at Exercise 9.2 in the Workbook for an analysis of how this affects efficiency and the distribution of economic surplus. But there are also markets with price ceilings. Rental prices in large cities in Europe are often regulated by the authorities, with prices set below the free-market level (that's the whole point—to avoid excessive housing costs). Consider Fig. 10.9, where Q is the number of rental units, and S is supply while D is demand. The market equilibrium would occur at point a, with rent P_a (think of it as the price per square metre), but the authorities have decided that the rent must not exceed P_{low}.

At that price, demand is Q_d while supply is only Q_{low}. This creates a queue of people who would like housing at this low price but can't get it.

Assume for simplicity that the number of people in the queue equals the number who get rent-controlled housing. In the figure, this means Q_{low} is half of Q_d. Imagine that the queue by chance is organised so that only those with the highest willingness to pay get housing. In that case, consumer surplus is area ABC, while producer surplus is area B. The deadweight loss is marked as †, reflecting the reduced number of rental units caused by the price ceiling.

As the polar opposite, imagine now that those with the *lowest* willingness to pay get housing first—those along the line segment from c to d on the demand

Fig. 10.9 Price regulation. *Note* The market equilibrium is at point a, but a price ceiling of P_{low} results in a supplied quantity of Q_{low}. At that price, demand is Q_d, so not everyone who wants housing will find it. The deadweight loss † is caused by too few units being supplied at this regulated price. In addition, there is a loss of efficiency if people with lower willingness to pay get ahead of those with higher willingness to pay in the queue

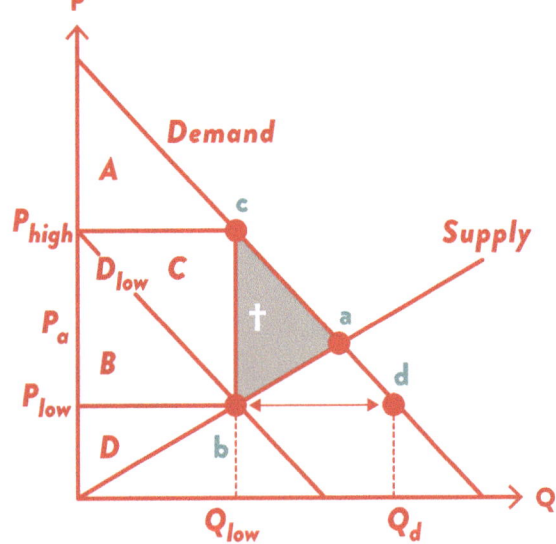

curve D. Demand from this group is shown as D_{low}, and now consumer surplus is only area B. Areas A and C are lost—an efficiency loss caused by the "wrong" people getting housing. The problem with price regulation is therefore both that too few units are supplied and that those who benefit are not necessarily the ones who value the housing most.

It's hard to fight market forces, and price regulations often lead to creative ways to jump the queue. In cities with rental control, we often find people paying large sums under the table to the owner, sometimes with agents as intermediates, to secure a flat. While illegal, these side payments improve economic efficiency by sorting the queue by willingness to pay. But of course, one may object to the distributional effects and the ethics of such a system, as it sustains intermediaries and normalises dishonest behaviour. Even in countries without rent controls, soaring housing costs have led some to question the fully market-based approach. Microeconomics gives you a framework to think clearly about housing policies. Once again, I hope you see the practical value of the theory!

10.8   Summary

Taxes are necessary for the functioning of society—to provide publicly provided goods like education, infrastructure, and defence, and to support those in need. Taxes can also be used to reduce pollution and thereby correct a market failure. Taxation is also one of the most hotly debated issues in politics and a clear dividing line between the political left and right. One reason for this is that taxes affect both economic efficiency (usually creating deadweight losses) and distribution (who bears the cost—consumers or producers?).

We have also looked at other types of market intervention, such as quotas (e.g., CO_2 permits) and price regulation (e.g., rent control). Just like taxes, these also have implications for both efficiency and distribution. What we've learned about taxation therefore has clear relevance for understanding other forms of economic policy.

10.9   Key Terms

Tax incidence: who bears the cost of a tax, consumers or producers.
Producer price: the price received by producers.
Consumer price: the price that consumers pay.
Tax wedge: the gap between the price paid by consumers and that received by producers.
Laffer curve: illustrates the (hump-shaped) relationship between the tax rate and total tax revenue collected by the government.
Tariff: tax on imported goods.
Pigouvian tax: tax on activity that creates a negative externality, such as pollution.

Emssion permits: an allowance that gives a company the legal right to emit one unit (e.g., one tonne) of a pollutant, usually carbon dioxide (CO_2).

Quota: a restriction on the amount of a specific good that can be sold in a market.

Price regulation: government determined prices in a market.

10.10   Multiple-Choice Exercises

10.1: Who Benefits from the Consumption Subsidy?
Imagine a market described by the figure below, where the initial equilibrium is at point a. The authorities then introduce a subsidy s per unit consumed. Who benefits from this subsidy?

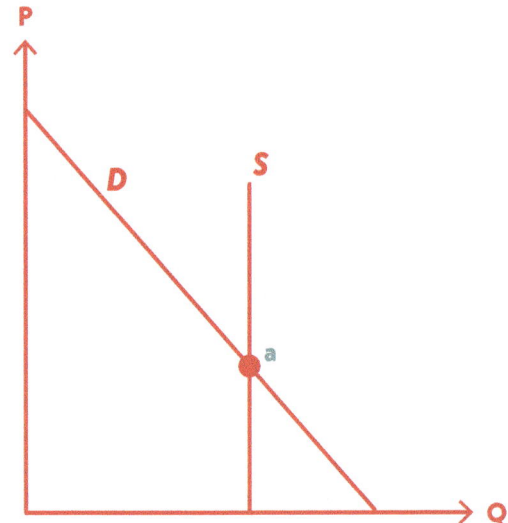

A. The subsidy only increases producer surplus
B. The subsidy only increases consumer surplus
C. The subsidy gives an equal increase in producer and consumer surplus
D. The subsidy affects neither consumer nor producer surplus

10.2: Tax and Deadweight Loss
Look at the figure below. Initially, equilibrium is given by point a in both markets. The authorities then introduce the same production tax t in both markets. In which market does this tax create the greatest deadweight loss?

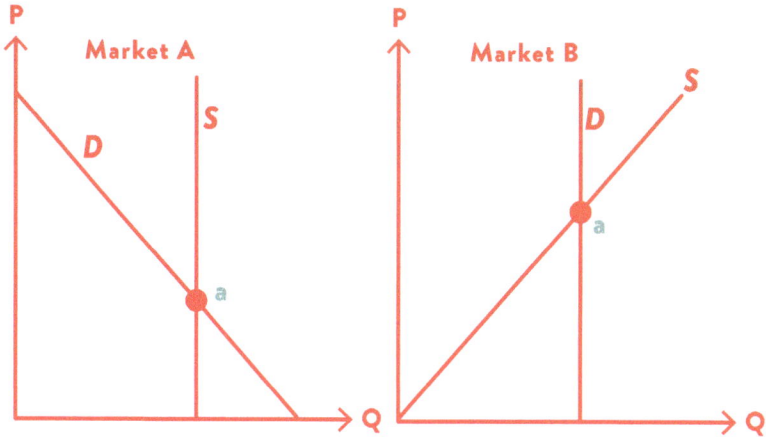

A. No deadweight loss is created in either market
B. Largest deadweight loss in market A
C. Largest deadweight loss in market B
D. Same positive deadweight loss in both markets

10.3: Positive Externality
Imagine a market as illustrated in the figure below, with equlibrium at point a. Assume that there is a positive external effect from production, which calls for economic policy to improve economic efficiency.

With optimal policy, which of the following statements is true?

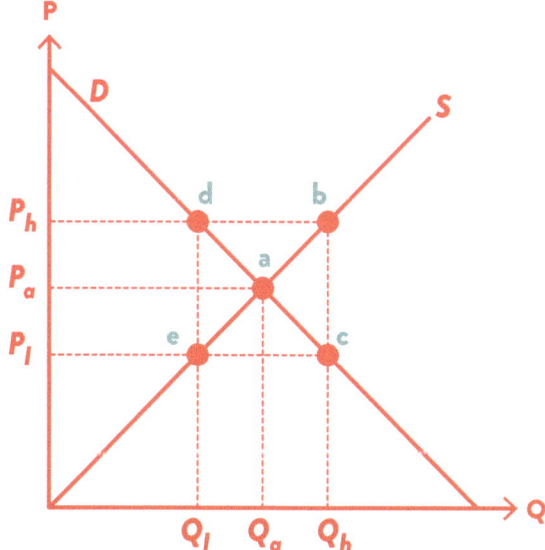

With optimal policy, we get:

A. Quantity Q_h, producer price P_h, consumer price P_l
B. Quantity Q_h, producer price P_l, consumer price P_h
C. Quantity Q_l, producer price P_h, consumer price P_l
D. Quantity Q_l, producer price P_l, consumer price P_h

10.4: Quantity Restriction

To engage in fish farming, one needs a permit from the authorities, a so-called concession. The willingness to pay for engaging in fish farming is given by the demand curve, with choke price P_D^{choke} and with demand Q_b if entry were completely free.

Assume that the number of concessions is given by $S = Q_c$. If these concessions were sold to the highest bidder, the market price would have been P_a. Now, a proposal has been made to instead distribute the concessions through a lottery, where everyone has an equal chance of entering the market. The demand curve (for those who have won the lottery) is now given by Demand'. What is the deadweight loss of such a solution compared to the market solution?

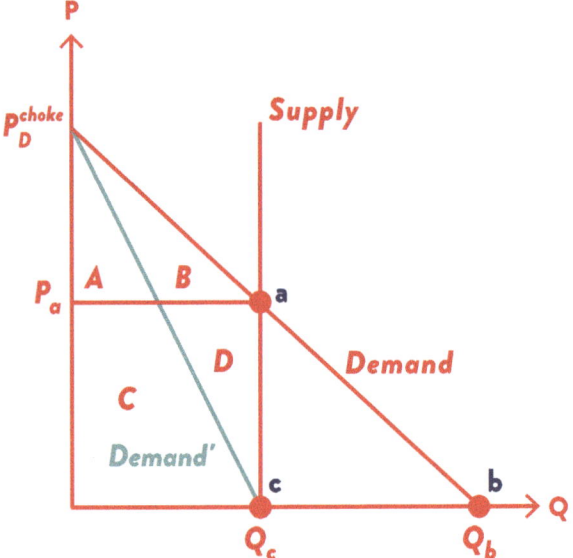

A. $A + B$
B. $A + C$
C. B
D. $B + D$

Solutions: 10.1 A; 10.2 A; 10.3 A; 10.4 D

Part IV
Market Power and Strategy

Conrad is on a roll. He has started his own publishing company and has also begun producing cardboard. However, these new ventures bring new challenges. While Conrad operates in a perfectly competitive market when producing A4 paper, he has market power in the sale of books and cardboard. This means he can no longer take the price for granted. Consequently, he must study the demand curve carefully. Additionally, if there are competitors, he will also need to think strategically about how to deal with them.

Chapter 11 is about monopoly—markets where there is only one producer. Conrad's publishing house has released a new microeconomics textbook and holds a monopoly on this product. What price should they set for the textbook?

Chapter 12 analyses oligopoly markets, where there are a few suppliers. For example, if there are two, this is called a duopoly. In the cardboard market, Conrad faces competition from an established producer making an almost identical product. How much cardboard should Conrad produce? The answer depends on how many the competitor produces, and since the competitor thinks in the same way, a strategic game develops between the two. This chapter also explores what happens when a new microeconomics textbook enters the market. How will this affect the price Conrad sets for his book? We use these stories to go through oligopoly models with both quantity and price as strategic variables, where firms are either equals or where one leads and the other follows.

Chapter 13 introduces game theory, a flexible tool for analysing a wide range of strategic situations—not just pricing and quantity decisions, but also everyday situations involving conflict.

This is going to be exciting!

Monopoly

11

Conrad has started a publishing house and is about to release a new microeconomics textbook. But how many copies should he produce, and what price should he choose?

Anna tries to convince Conrad that elasticities are incredibly important for a monopolist.

11.1 Introduction

Conrad has a passion for paper. Everything made of paper, not the least books. And so, he's started his own publishing house, Alpha Books. The first thing he does as a publisher is to launch a book on microeconomics, a simple introductory textbook written by Anna and Brian's lecturer at the School of Economics, and he simply refers to it as the ABC. Alpha Books is the only publisher with the rights to produce this book: Alpha Books has a monopoly on the ABC.

Conrad has started his own publishing house

What price should the monopolist set for the microeconomics textbook? How strong is Alpha Books' market power, really? And if the monopolist can price discriminate between different customer groups, which prices should it set? Finally, why is monopoly considered problematic from a societal perspective, and can monopolies sometimes be beneficial? We will answer all of this in this chapter!

We start by introducing the key concept of marginal revenue and then derive the monopolist's optimal choice of quantity. Even though the monopolist is the only producer of this product, there may be substitutes, which limits how high the monopolist can set the price: the price elasticity of demand indicates how strong the monopolist's market power is. We then move on to the question of price discrimination before studying the societal implications of monopoly, both regarding efficiency and distribution. Finally, we address the case of natural monopolies—situations where monopoly can be justified from an economic efficiency perspective.

11.2   Marginal Revenue

While the price of A4 paper is set by the market and outside Conrad's control, the situation is a bit different for the microeconomics textbook. Alpha Books is the only producer of the ABC and Conrad realises he has influence over the price.

He goes to see Anna, whom Conrad has reassigned as sales manager at the publishing house. Anna explains to her grandfather that he needs to think about marginal revenue—this is in fact a central topic in the monopoly chapter of the textbook, which she recommends him to read. Marginal revenue expresses the change in revenue from producing one more unit of the good.

Mathematically, we can find this by starting from revenue (R), which can be expressed as:

$$R(Q) = P(Q)Q \quad \text{Revenue} \tag{11.1}$$

Here, P is the price per book, and Q is the number of books, and we see that the monopolist takes into consideration that the price is a function of the quantity produced, that is, $P(Q)$. The marginal revenue, abbreviated MR, is the derivative of revenue with respect to quantity:

$$MR = \frac{\partial R}{\partial Q} = P + Q\frac{\partial P}{\partial Q} \quad \text{Marginal revenue} \tag{11.2}$$

The first term on the right side shows that selling one extra unit generates revenue equal to the price, while the second term accounts for the fact that an increase in quantity leads to a lower price, which applies to all units sold.

Note the difference between a monopoly firm and a perfectly competitive firm: In perfect competition, firms take prices as given, so $(\partial P/\partial Q) = 0$, which means the second term on the right side of (11.2) is zero, and marginal revenue equals the price.

In contrast, a firm with market power understands that increased production lowers the price, $(\partial P/\partial Q) < 0$. When the monopolist decides how much to produce—and thus the price—it is crucial to understand exactly how much the price falls when quantity increases, or how much the price rises if production is reduced.

Conrad wonders what determines the price sensitivity of demand, and Anna explains that it depends on what alternatives customers have and how good these alternatives are. Monopoly does not mean a total absence of competition: there are competing textbooks on the market, students can borrow the book from the library, from each other, or buy used copies.

For example, if students at the School of Economics, where the ABC is used, think all microeconomics books are pretty much the same, the demand curve will be quite flat: we say demand is elastic. On the other hand, if students consider the ABC as quite unique, the demand curve could be quite steep, and we say demand is less elastic.

Let's go through an example of marginal revenue based on the linear demand function $Q = \alpha - \beta P$. We have already seen that this can be rewritten in inverse form as $P = (\alpha - Q)/\beta$.

We can differentiate the inverse demand with respect to Q to find the slope of the demand curve:

$$\frac{\partial P}{\partial Q} = -\frac{1}{\beta} \quad \text{Slope of the demand curve} \qquad (11.3)$$

Insert this in the expression for the marginal revenue and we find:

$$MR = \frac{\partial R}{\partial Q} = P + Q\frac{\partial P}{\partial Q} = \frac{\alpha - Q}{\beta} - \frac{Q}{\beta} = \frac{\alpha - 2Q}{\beta} \quad \text{Marginal revenue} \quad (11.4)$$

The slope of the marginal revenue curve is given by:

$$\frac{\partial MR}{\partial Q} = -\frac{2}{\beta} \quad \text{Slope of marginal revenue curve} \qquad (11.5)$$

We observe that with a linear demand curve, marginal revenue falls twice as fast as the demand curve. The quantity that gives zero marginal revenue is:

$$MR = 0 \Rightarrow \frac{\alpha - 2Q}{\beta} = 0 \Rightarrow Q = 0.5\alpha \quad \text{Marginal revenue equals zero} \qquad (11.6)$$

Does this look familiar? Check Math Box 8.1. This is exactly the level of Q that makes demand unit elastic ($\varepsilon = 1$)! It also means that when $MR > 0$, demand is elastic ($\varepsilon > 1$), and when $MR < 0$, demand is inelastic ($\varepsilon < 1$). The relationship

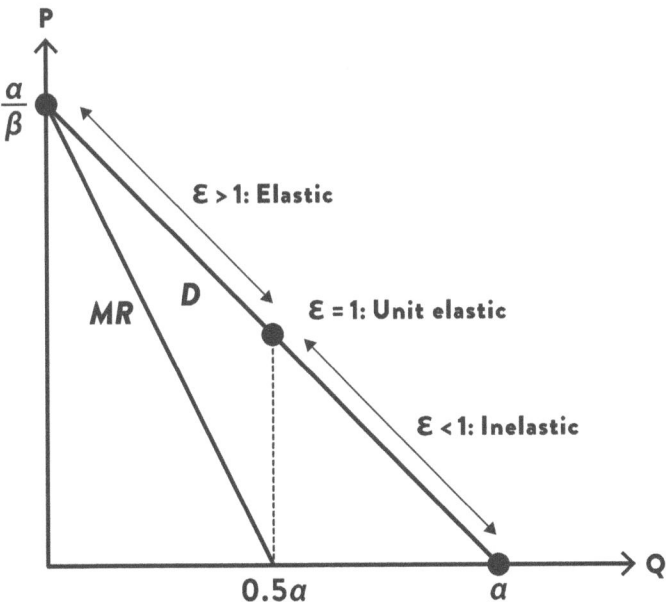

Fig. 11.1 Demand and Marginal Revenue. *Note* The marginal revenue curve (MR) shows the change in revenue when production increases by one unit. It falls twice as fast as the demand curve. The reason marginal revenue falls faster than the demand curve is that an increase in production lowers the willingness to pay, resulting in a lower price for all units sold. Notice that the marginal revenue curve crosses the horizontal axis at the point $Q = 0.5\alpha$, which is where demand is unit elastic

between the demand curve, marginal revenue, and price elasticity of demand is illustrated in Fig. 11.1. Notice that the marginal revenue curve falls twice as fast as the demand curve and therefore crosses the horizontal axis at the point $Q = 0.5\alpha$.

From the discussion above, summarised in Fig. 11.1, we see that there is a close relationship between marginal revenue and price elasticity of demand. Let me describe this relationship mathematically. We start by rewriting the expression for marginal revenue by multiplying the second term by the price P both in the numerator and denominator:

$$MR = P + Q\frac{\partial P}{\partial Q} = P + P\frac{Q}{P}\frac{\partial P}{\partial Q} = P\left(1 + \frac{Q}{P}\frac{\partial P}{\partial Q}\right) \qquad (11.7)$$

We see that the derivative expression in (11.7) resembles the price elasticity of demand, but not exactly! We need to invert it. And if we also add a minus sign, the denominator will be equal to the price elasticity of demand:

$$\frac{Q}{P}\frac{\partial P}{\partial Q} = \frac{1}{\frac{\partial Q}{\partial P}\frac{P}{Q}} = -\frac{1}{\varepsilon}$$

This means that (11.7) can be written as:

$$MR = P\left(1 - \frac{1}{\varepsilon}\right) \tag{11.8}$$

A useful insight from this expression is that a monopolist will never produce where demand is inelastic. This is clear because with inelastic demand ($\varepsilon < 1$) we have $MR < 0$, which means that revenues go up with a reduction in supply. And since a reduction in supply also leads to cost savings, this must necessarily be profitable. This insight will be central as we now discuss the monopolist's optimal choice of quantity, and thus price.

11.3   Profit Maximisation

Anna has now given Conrad a thorough introduction to demand. The preliminary conclusion is that one should produce where demand is elastic. But let's find out more precisely which price the monopolist should choose to maximise its profit π, defined as revenue $R(Q)$ minus costs $C(Q)$:

$$\pi = R(Q) - C(Q) \quad \text{Profit} \tag{11.9}$$

We differentiate profit with respect to quantity to find the first-order condition for the optimal quantity:

$$\frac{\partial \pi}{\partial Q} = \frac{\partial R}{\partial Q} - \frac{\partial C}{\partial Q} = 0 \quad \text{Profit maximisation} \tag{11.10}$$

The first term on the right-hand side is the change in revenue resulting from a marginal increase in the quantity produced, that is, the marginal revenue, while the second term is the change in cost resulting from a marginal increase in quantity produced: the marginal cost. We can therefore write the first-order condition as:

$$MR = MC \quad \text{Optimal production rule} \tag{11.11}$$

To maximise profit, the firm should choose a quantity where marginal revenue equals marginal cost. This resembles the condition for optimal production in perfect competition, $P = MC$, except that for a perfectly competitive firm, marginal revenue equals the price, while for a monopolist, it's more complicated since the price depends on quantity.

Throughout this part of the book, we simplify the cost side by assuming constant marginal cost. We can think of firms as having a Leontief technology, meaning marginal cost is constant up to a certain capacity limit (as discussed in

Sect. 6.3, where marginal cost becomes a step function), and where the capacity constraint is not binding for the production interval under consideration.

As for Alpha Books, they don't print the books themselves but outsource this to several printing companies. Since the printing industry is perfectly competitive, these printers supply the book to the publisher at marginal cost. We assume there is ample production capacity in the printing sector, so marginal cost is constant.

The reason for simplifying the cost side is to focus on demand: both the monopolist and you, the student, have plenty to consider here! But note that this is only a simplification—the analysis applies regardless of the shape of the marginal cost curve.

Figure 11.2 builds on Fig. 11.1 by adding marginal costs. The optimal production level is where MR crosses MC, giving an equilibrium on the demand curve at point m, with quantity Q_{Mon} and price P_{Mon}. Here, we also compare the monopolist's choice to the perfect competition solution, given by point f where price equals marginal cost, yielding Q_{Comp} and P_{Comp}.

We see that the monopolist produces less and charges a higher price than in perfect competition. The fact that the monopolist sets a price above marginal cost reflects market power. The greater the gap between price and marginal cost, the more market power the monopolist has.

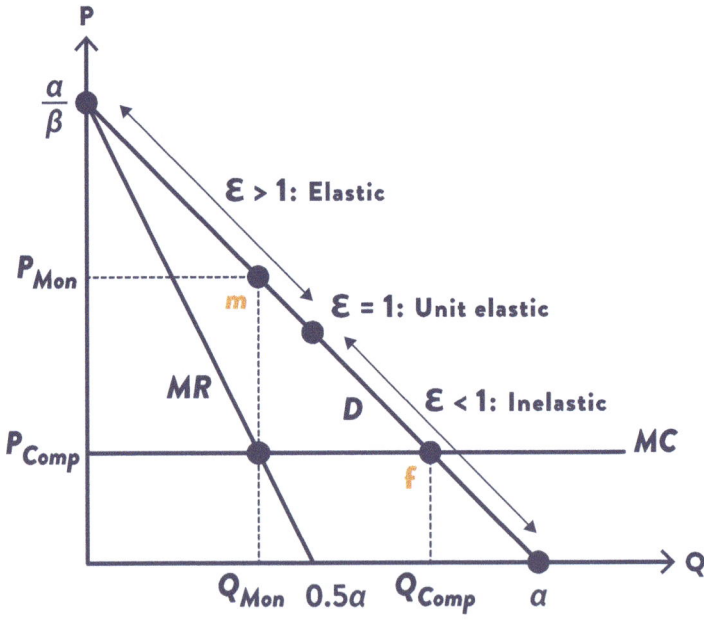

Fig. 11.2 The monopolist's choice. *Note* The monopolist chooses the quantity so that marginal revenue equals marginal cost. In the figure, we see this gives quantity Q_{Mon} and price P_{Mon}. This means a higher price and lower quantity than the perfect competition outcome, shown by P_{Comp} and Q_{Comp}

In Math Box 11.1, we demonstrate the monopolist's choice using a linear demand function.

Math Box 11.1: Monopolist's Choice

Assume that demand is given by:

$$Q = \alpha - \beta P$$

This can be written in inverse form as:

$$P = \frac{\alpha - Q}{\beta}$$

This gives the monopolist a revenue, price multiplied by quantity produced:

$$R(Q) = \left(\frac{\alpha - Q}{\beta} \right) Q \quad \text{Revenue}$$

The marginal revenue is then given by:

$$MR = \frac{\partial R}{\partial Q} = \frac{\alpha - 2Q}{\beta} \quad \text{Marginal revenue}$$

Assume that the cost function is given by:

$$C(Q) = cQ \quad \text{Costfunction}$$

This means that the marginal cost is constant:

$$MC = \frac{\partial C}{\partial Q} = c \quad \text{Marginal cost}$$

Maximising profits gives us:

$$MR = MC \Rightarrow \frac{\alpha - 2Q}{\beta} = c \quad \text{Profit maximising condition}$$

This means that the monopolist chooses a quantity:

$$Q_{Mon} = \frac{\alpha - \beta c}{2} \quad \text{Monopolist's quantity}$$

And the monopoly price becomes:

$$P_{Mon} = \frac{\alpha - Q_{Mon}}{\beta} = \frac{1}{2}\left(\frac{\alpha}{\beta} + c\right) \quad \text{Monopoly price, the midprice rule}$$

We see that the monopolist will choose a price that lies halfway between the choke price α/β and the marginal cost (c). We shall call this *the midprice rule*, and it applies in the case of constant marginal costs.

11.4   How Much Market Power Does the Monopolist Really Have?

What determines the market power of a monopolist? Once again, we will see that the price elasticity of demand is important.

To understand the relationship between monopoly power and demand elasticity, let's use the relationship between marginal revenue and demand elasticity that we established above, in Eq. 11.8, along with the optimality condition $MR = MC$, so that we can write:

$$MR = P\left(1 - \frac{1}{\varepsilon}\right) = MC \tag{11.12}$$

This can be expressed as:

$$P - MC = \frac{P}{\varepsilon}$$

Which in turn can be formulated as:

$$\frac{P - MC}{P} = \frac{1}{\varepsilon} \quad \text{Lerner index} \tag{11.13}$$

The expression on the left side is the Lerner index, named after the Russian-born economist Abba Lerner (1903–1982). It measures market power as the gap between the monopoly price and marginal cost (which would define the price under perfect competition) expressed as a percentage of the monopoly price. We see that a monopolist will choose a markup such that the Lerner index is inversely proportional to the price elasticity of demand, ε. The higher the ε—that is, the more elastic the demand—the lower is the markup, which implies lower market power. This makes sense: the higher the demand elasticity, the more people stop buying the good when the price increases, and the less power the monopolist has over consumers. Conversely, a low demand elasticity means the monopolist does not risk losing many customers when the price rises: his power over consumers is high, resulting in a high price.

11.5   Demand Elasticity and Monopoly Power

The School of Economics has admitted a larger cohort than the previous year, creating the opportunity for increased sales of the textbook in microeconomics published by Alpha Books. However, the new students can buy the book second-hand from last year's students (who have decided the book was not worth keeping!). Last year's cohort had no option to buy second-hand, as the book had just been released.

Anna discusses with Conrad what price they should set for the textbook this year. A recent market survey Anna conducted shows that they should be able to sell an equal number of copies as before at the original price, but Anna is considering whether there is a reason to change the price.

Anna and Conrad discuss the price of the textbook in microeconomics

In Fig. 11.3, she draws last year's demand for the book as D_1 and this year's demand as D_2. Anna thinks that if they increase the price, many students will choose the second-hand option. This means that a price increase this year will cause a larger drop in demand than a similar price increase would have caused last year when the book was new and therefore second-hand was not an option. On the other hand, if they lower the price, they can sell more books because the cohort has grown.

The demand elasticity of the new cohort is higher than that of the old one, and Anna argues that Alpha Books should now choose a quantity where MR_2 crosses MC, which is point n on D_2, meaning a lower price for the book. A more elastic demand thus leads to a lower price, consistent with the Lerner index. One can say

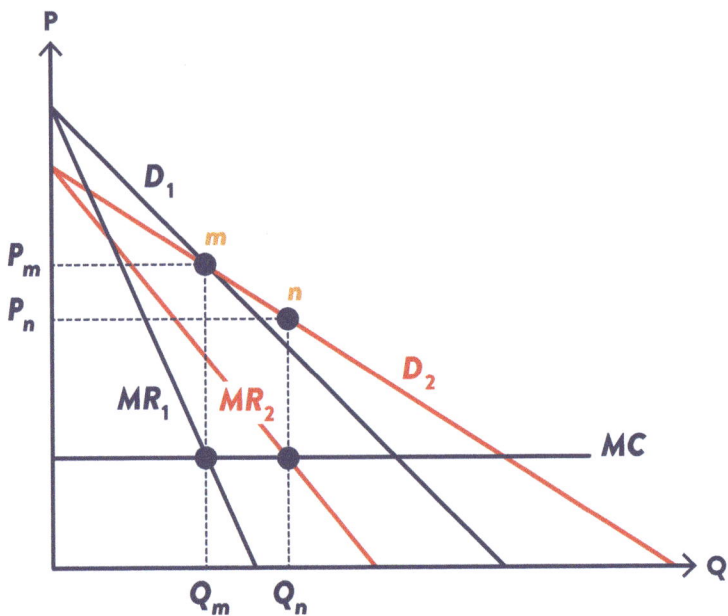

Fig. 11.3 Monopolist's choice with increased demand elasticity. *Note* Initially, demand is given by D_1, and the monopolist's optimal choice is at point m. The demand curve D_2 represents an increase in demand elasticity, leading to a new optimum at point n. We see that the higher demand elasticity has resulted in a lower price

that the monopolist has lost some of its market power because this year's students can alternatively buy the book second-hand.

11.6 Price Discrimination

So far, we have assumed that the monopolist sets a single price. That is, we have ignored the possibility of price discrimination. There are many types of price discrimination and many clever tricks a monopolist can use to extract as much surplus as possible from consumers. But what they all have in common is that the monopolist wants to charge a high price to consumers with a high willingness to pay and a low price to those with a low willingness to pay.

We will keep to a brief presentation of the main intuition here and leave a more detailed analysis of price discrimination to advanced microeconomics courses.

Let's say the School of Economics has started an executive program aimed at professionals seeking further education. One of the courses offered is an introductory microeconomics course, for those who do not already have this as part of their education, and the syllabus naturally includes the ABC.

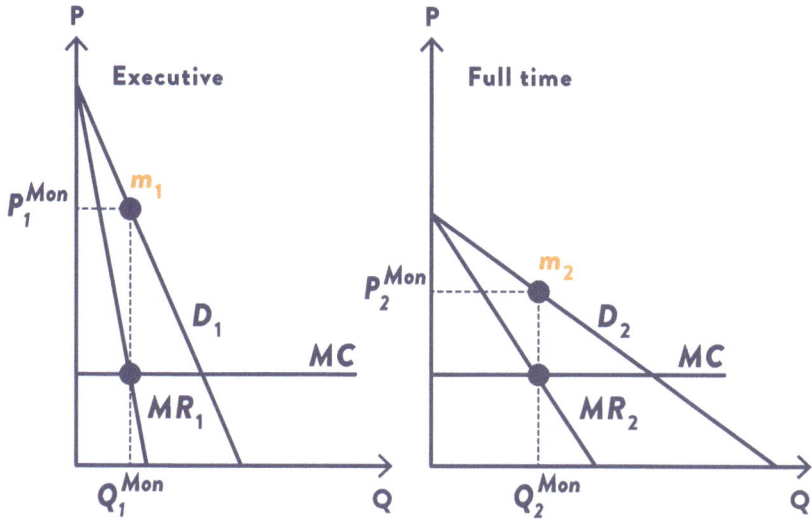

Fig. 11.4 Price discrimination. *Note* The figure on the left shows the monopolist's choice when willingness to pay is high and price sensitivity is low (D_1, executives), while the figure on the right shows the monopolist's choice when willingness to pay is lower and price sensitivity is higher (D_2, full-time students). We see that the executive students must pay a higher price than the full-time students, $P_1^{Mon} > P_2^{Mon}$

Since these are working adults, whose employers may even cover their cost, it is likely that they both have a higher willingness to pay and a lower price sensitivity in their demand for learning materials than regular full-time students.

Anna sits down and thinks about how Alpha Books should relate to this new customer group. She draws Fig. 11.4, where the left side shows demand in the executive market, and the right side shows demand among full-time students.

The analysis clearly shows that Alpha Books should set a higher price for the executive students than for the full-time students. The optimal choice in the executive market is at point m_1 with price P_1^{Mon}, while for the full-time students it is at point m_2 with the lower price P_2^{Mon}.

But how should this differential pricing be implemented in practice? One option could be that executive students pay full price, while full-time students receive a discount upon showing proof of paid tuition fees. Good idea! But it is important that full-time students cannot resell the books to executive students. However, Anna does not think this will be a big problem, since the two groups rarely or never meet (the executive program is based on weekend gatherings at rather fancy hotels, and the older students are not particularly interested in buying cheap books online since their employer pays anyway). She decides to go straight to Conrad and present the proposal!

11.7   Deadweight Loss from Monopoly

Monopoly doesn't sound good. There are two reasons for this. First, the monopolist charges a high price, which harms consumers. This is a distributional issue. Second, monopoly leads to an efficiency loss.

Look at Fig. 11.5, and let's start with the consumers. Compared to the perfect competition solution at point f, monopoly results in the equilibrium at point m with lower quantity and higher price. Under perfect competition, consumer surplus is the entire area ABC under the demand curve and above the marginal cost. With monopoly, consumer surplus is reduced to area A above the monopoly price. Consumers thus lose the area BC.

The producer captures some of the surplus from consumers, but not all of it. We see that producer surplus (which was zero under perfect competition) is now given by area B, but this means there is an area C that consumers lose, but which is not captured by the producers. As discussed in the previous chapter, this outcome is not optimal from an economic efficiency perspective because the market solution does not maximise the sum of consumer and producer surplus. The triangle C is the efficiency loss, a deadweight loss.

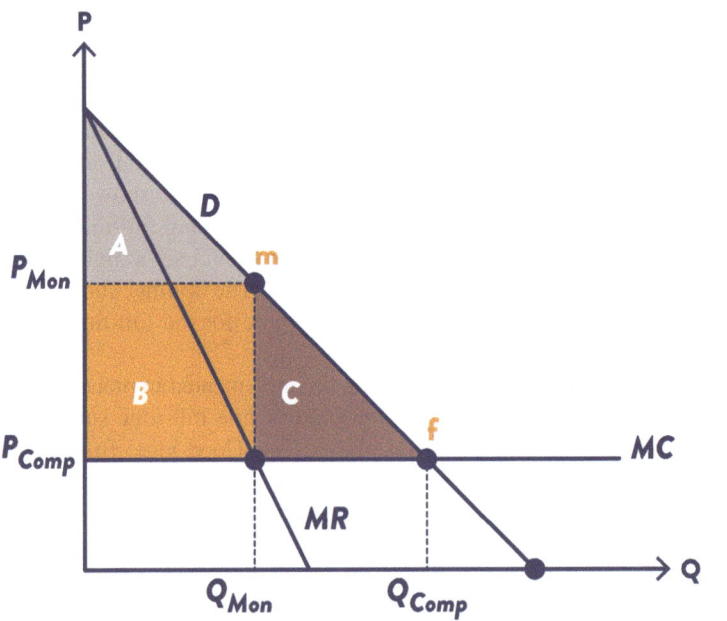

Fig. 11.5 Deadweight loss from monopoly. *Note* Monopoly causes a deadweight loss to society. The perfect competition solution is at point f, where consumer surplus is given by area ABC. The monopoly equilibrium is at point m, where producer surplus is area B and consumer surplus is reduced to area A. Area C represents the deadweight loss to society caused by the monopoly

11.8   Natural Monopoly

Alpha Books has a monopoly on its textbook in microeconomics. Other publishers can produce similar textbooks, but not one identical to the ABC. It is protected by copyright. But why do authorities grant this type of monopoly protection? We have just shown that monopoly creates an efficiency loss compared to perfect competition!

Some markets are so-called *natural monopolies*. Natural because the cost structure makes perfect competition impossible or undesirable. The market for books is such a market. Writing a book is time-consuming and thus constitutes a large, fixed cost, while the marginal cost of printing the books is constant. This means total average costs are falling. If anyone were allowed to print the ABC, competition would drive the price down to marginal cost, and the operating profit would be zero: revenue from sales would cover the printing costs, but nothing would remain to cover the fixed costs. It's easy to see that not many books would be written in such a market!

Similarly, pharmaceutical companies invest large sums in research and development of a new vaccine during a pandemic, but once the medicine is developed and approved, the cost of production is very low. If competitors were free to copy the medicine, pharmaceutical companies would have no incentive to develop it in the first place. Authorities can in such cases allow a monopoly, for example in the form of copyright on a book or patents on a medicine.

Other examples of natural monopolies are markets that require large physical infrastructure, such as electricity through the high-voltage grid or transport by rail, where there are large fixed costs to build the infrastructure, but once the investment is made, the variable costs are quite low (it costs little to send electricity through the high-voltage grid and little to run trains on the tracks). In these cases, the infrastructure is often state-owned monopolies.

Figure 11.6 shows a natural monopoly, with declining total average costs. A monopolist will choose the equilibrium at point m, with quantity Q_{Mon} and price P_{Mon}. Since this price is higher than average costs, the firm makes a profit: the book will be written, and the medicines developed.

But are there other ways to manage natural monopolies besides simply granting monopoly rights? The ideal, at least in theory, is to set a price ceiling equal to marginal cost, resulting in the equilibrium at point f, and then fully compensate the producer for fixed costs. But besides the problem that authorities may not know the firm's costs, there is the challenge that the state must finance the compensation somehow, usually through taxation. This can itself be costly and not always popular politically.

An alternative is to grant the firm monopoly rights but regulate the price, such that it can cover its costs, but no more. This means a price ceiling P_g, which leads to consumer demand Q_g. The challenge for authorities is that they do not necessarily know the firm's cost structure, making it difficult to set the right price.

In summary, we can say that the monopoly solution at m, protected by copyright and patents, can be a pragmatic way to solve the problem of natural monopolies

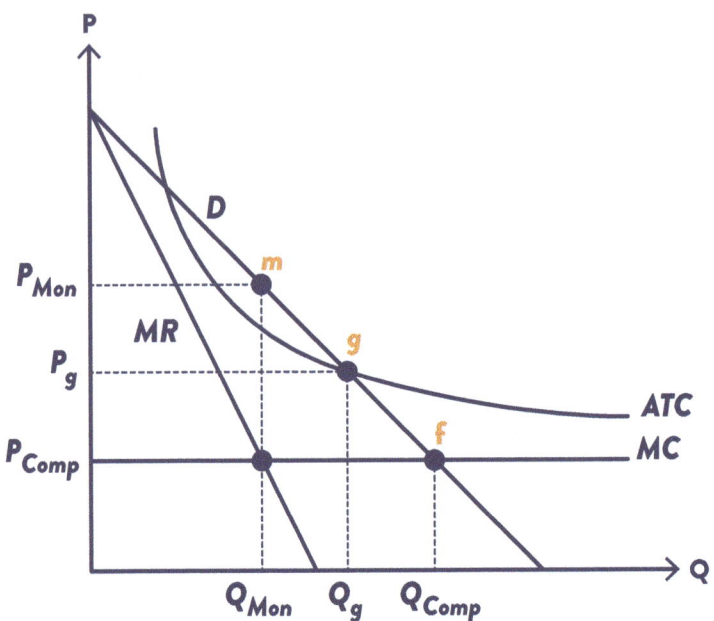

Fig. 11.6 Natural monopoly. *Note* A natural monopoly is characterised by declining total average costs. Under perfect competition, the fixed costs would not be covered, making investment unprofitable. To ensure innovation and investment in infrastructure, authorities can grant monopoly rights in such markets. The monopolist will choose the equilibrium at point *m*, which yields positive profit. Alternatively, authorities can set a price ceiling P_g that allows the firm to cover its costs but no more. The socially optimal solution would be to set a price ceiling at P_{Comp} and then cover the fixed costs through a subsidy

when price regulation is difficult and when state ownership is not on the political agenda.

11.9   Summary

In this chapter, we have seen that the monopolist must have a profound understanding of the demand side since, unlike in perfect competition, the price cannot be taken as given. Two key concepts describe demand: price elasticity and marginal revenue. We examined each concept individually and how they relate to each other.

We then explored how the monopolist maximises profit by choosing a quantity where marginal revenue equals marginal cost. The price in a monopoly market is higher than under perfect competition. We also studied price discrimination and saw that a monopolist will charge higher prices to customer groups with a high willingness to pay and low price sensitivity.

Compared to perfect competition, the monopolist captures some of the consumer surplus by restricting production. This creates a distribution issue and results in deadweight loss, an economically inefficient outcome.

Finally, we saw that one important reason why monopolies can sometimes be justified from a societal perspective is when total average costs are declining—so-called natural monopolies. In such cases, perfect competition is not feasible because the price would be too low to cover fixed costs, and firms would have no incentive to innovate or invest in necessary infrastructure.

11.10 Key Terms

Marginal revenue: The change in revenue from producing one additional unit.
Price discrimination: When a firm with market power charges different prices to different types of customers.
Natural monopoly: When production results in declining total average costs throughout the entire relevant production range.

11.11 Multiple-Choice Exercises

11.1: Elasticities
Based on the figure below, at what quantity is the demand unit elastic?

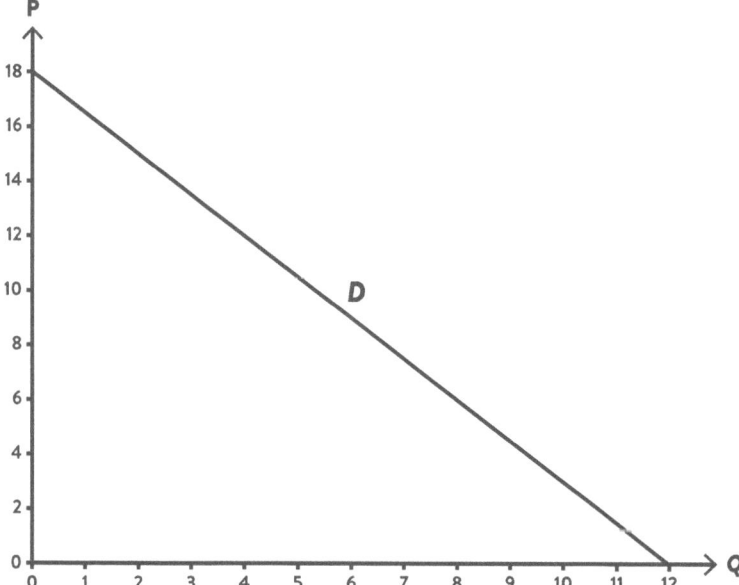

A. $Q = 3$
B. $Q = 6$
C. $Q = 12$
D. $Q = 18$

11.2: Marginal Revenue

Using the same figure as above (in 11.1), what is the functional expression for the monopolist's marginal revenue?

A. $MR = 6$
B. $MR = 18 - 1.5Q$
C. $MR = 18 - 3Q$
D. $MR = 12 - 0.5Q$

11.3: Monopoly Profit

Assume $MR = 18 - 3Q$ and $MC = 6$. What is the monopoly profit in this market?

A. 18
B. 20
C. 22
D. 24

11.4: Market Power

Which of the following four statements is correct?

1. The Lerner index is zero under perfect competition
2. The Lerner index is higher for higher demand elasticity
3. If the demand elasticity is 1.60, the Lerner index is 0.625
4. If $P = 100$ and marginal cost is 40, the Lerner index will be 0.6

A. 1,2 and 3
B. 1, 3 and 4
C. 3 and 4
D. 4

Solutions: 11.1 B; 11.2 C; 11.3 D; 11.4 B

Oligopoly

12

Conrad is on a roll and has decided to start producing cardboard, challenging the established producer Firstpak. How will this play out?

Things are heating up in the book market too. A competing publisher has released a new microeconomics textbook—how should Alpha Books respond to this new situation?

12.1 Introduction

In the market for A4 paper, Conrad is just one of many producers. A4 is a standard-ised product, and it doesn't require major investments in research and development. Life as a producer in such a market isn't particularly complicated. If you keep your costs under control, you simply produce up to the point where the cost of making the last sheet equals the market price.

Conrad admires his newly developed milk cartons

In contrast, as the publisher of a new textbook in microeconomics, Alpha Books enjoys a monopoly position. A monopolist has the power to set a price above marginal cost, but this power is not unlimited: if the price is too high, customers may switch to another book or buy a used one. Life as a monopolist isn't especially complicated either, but it does involve more considerations than under perfect competition. Not only must you keep an eye on your costs, you also need to think carefully about the demand side to find the optimal price.

Life as an oligopolist is much more complex, and there isn't one theory to describe all situations. In this chapter, we will present the three main oligopoly models. But before we dive into the details of each one, we start with an overview that will also help you understand which model is most relevant in different settings.

12.2   Three Models of Oligopoly

The three models we cover in this chapter differ along two key dimensions: whether *quantity* or *price* is the strategic variable, and whether firms make their decisions *simultaneously* or *sequentially*.

We begin with two models where quantity is the strategic variable—this is called *quantity competition*. In one of these, firms choose quantities simultaneously; in the other, sequentially. Finally, we look at a model where *price* is the strategic variable and firms set prices simultaneously.

When firms compete in quantity and make their choices at the same time, we have Cournot competition, named after the French economist Augustin Cournot (1801–1877). When firms still compete in quantity, but one firm chooses before the other, we have *Stackelberg competition*, named after the German economist Heinrich Freiherr von Stackelberg (1905–1946). When price is the strategic variable and firms choose prices simultaneously, we have *Bertrand competition*, named after the French economist Joseph Bertrand (1822–1900).

What does it mean for quantity or price to be the strategic variable? Technically, this is the variable the firm chooses when maximising profit. In practice, *quantity* tends to be the strategic variable when firms face binding capacity constraints, whereas *price* is the strategic variable when there are no such constraints.

In the market for cardboard, for instance, once the capital equipment is in place, production is limited by installed capacity, and it takes time to expand that capacity. When firms choose quantity, we can think of this as setting production capacity.

By contrast, *price* is the strategic variable when production capacity is not an issue. Consider the publishing industry. Books can be printed in many different places. If demand for the ABC increases, Alpha Books can simply call up a printing press anywhere and have more copies produced at short notice. Capacity isn't a limiting factor here, and firms compete on price.

What about simultaneous versus sequential choices? In technical terms, simultaneous choice means that firms make their decisions at the same time. In practice, this means that a firm doesn't know what the competitor has chosen when it makes its own decision.

The alternative is *sequential choice*, where one firm moves first and the other follows.

As we will see in Stackelberg competition, the leader can gain a dominant position in the market by choosing a large output, knowing that the follower will respond by choosing a smaller quantity. In this case, the follower observes the leader's move before deciding—so choices are made sequentially.

12.3   The Cournot Model

Conrad is inspired by the energy brought to the firm by Anna and Brian. In fact, he feels younger than he has in years. His latest project is the production of cardboard. He has launched a dedicated division at the factory to focus on this new venture. The name is as simple as it is brilliant: Conrad's Box.

At Conrad's Box, or the Box for short, much effort has gone into designing the perfect product. Now they're ready to challenge Firstpak, a company which until now has enjoyed a monopoly in the market for cardboard, used in milk cartons and the like.

The next step is to invest in production equipment. This is a critical decision. The machines must be custom-built for each factory, and once installed, they're there to stay. Scaling capacity up or down is time-consuming—at least in the cardboard industry.

So, what production capacity should Conrad choose?

The answer depends on Firstpak's capacity. Conrad knows that Firstpak is in the process of replacing its old cardboard machines and is therefore in a similar position to Conrad. But he doesn't know what their plans are—just as Firstpak knows nothing of Conrad's intentions.

Right now, the teams at the Box and Firstpak are considering their capacity decisions independently. We can think of this as the two producers choosing capacity simultaneously. This is the classic Cournot duopoly.

Conrad speaks with Brian, who has been appointed to lead the new cardboard division. Brian has developed a solid understanding of economics, thanks in large part to his close collaboration with Anna in the paper production team. Conrad also has more personal reasons for bringing Brian into the management of the factory. He has seen how well Anna and Brian work together, and he's thinking about the future of the family business. It's important to him that the company stays in the family—even after Anna...

Brian points out that choosing the right quantity is a challenging question. He suggests tackling it step by step. He begins by analysing how the firm should respond to a given output level from the competitor. Then, he extends this to explore how they should react to different output levels from the competitor. Finally, he considers the situation from the competitor's perspective, to anticipate how they are likely to think about the problem, and what the market equilibrium might be. Based on this, he will be able to make a recommendation for what capacity Conrad should choose.

The demand for cardboard is given by the linear demand function:

$$Q = \alpha - \beta P \quad \text{Demand} \tag{12.1}$$

In inverse form, expressing price as a function of quantity, the demand can be written as follows:

$$P = \frac{\alpha - Q}{\beta} \quad \text{Inverse demand} \tag{12.2}$$

The total amount of cardboard supplied by the two firms—Firstpak, which we label A, and the Box, which we label B—is given by $Q = Q_A + Q_B$, and the inverse demand function can then be written as:

$$P = \frac{\alpha - Q_A - Q_B}{\beta} \tag{12.3}$$

Assume that Firstpak produces a given quantity $\overline{Q}_A$. What remains for Conrad's Box is called the residual demand and can be formulated as follows:

$$RD_B = \frac{(\alpha - \overline{Q}_A) - Q_B}{\beta} \quad \text{Residual demand for firm B} \tag{12.4}$$

Look at Fig. 12.1, and let's start with the simplest case. The figure on the left shows the situation where $\overline{Q}_A = 0$, meaning that firm B, the Box, is effectively a monopolist. The equilibrium is as we have seen before at point m, where the marginal revenue curve crosses the marginal cost.

If $\overline{Q}_A > 0$, then there is less demand left for the Box, as shown in the figure on the right. Here, the RD-curve is the residual demand, and MR_{RD} is the marginal revenue of the residual demand. The optimal choice for the Box is where MR_{RD} equals marginal cost, shown as point n in the figure. We see that the increased production by A (from zero to a positive amount) leads firm B to reduce its quantity, from Q_B^{Mon} to Q_B^n. We call this case strategic substitutes.

The analysis of residual demand shows that when Firstpak increases its quantity, Conrad's best response is to reduce its quantity. If Firstpak sets a sufficiently large quantity, Conrad will not produce at all. This will happen if the residual demand curve in Fig. 12.1 shifts so far inward that it intersects the vertical axis at or below the level of marginal cost. Firstpak will then have produced so much that the market is saturated: the price is too low for Conrad's Box to make a profit.

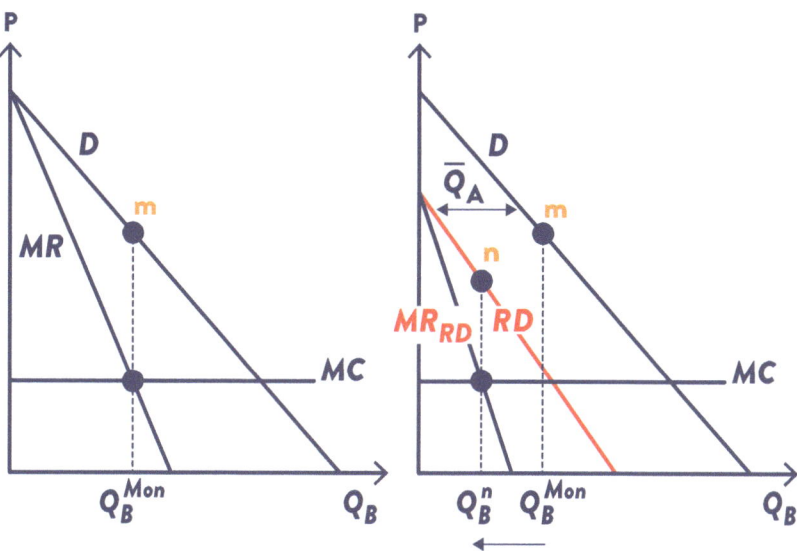

Fig. 12.1 Residual demand. *Note* The figure shows the residual demand (RD) facing firm B, which is the demand left for the firm once its competitor, producer A, has chosen its quantity. On the left, we see the situation where producer A has chosen zero quantity, so firm B is effectively a monopolist and produces Q_B^{Mon}. On the right, we see what happens when firm A chooses a positive quantity $\overline{Q}_A$, which shifts the demand curve facing firm B, i.e., the residual demand (RD), inward. The new marginal revenue curve for firm B is given by MR_{RD}. As a result, firm B's optimal production decreases from Q_B^{Mon} to Q_B^n

Figure 12.2 shows how the Box reacts to different quantities produced by its competitor. This is called firm B's reaction function: $R_B(Q_A)$ Producer B's profit falls along the reaction function. To see this, note that the reaction function starts at $Q_A = 0$, where B is a monopolist and chooses quantity Q_B^{Mon}. When A enters the market, B responds by reducing its quantity, and profit necessarily falls. In fact, if Q_A becomes large enough, B will not produce at all (where the reaction function crosses the horizontal axis), and without production there will, of course, be no profit.

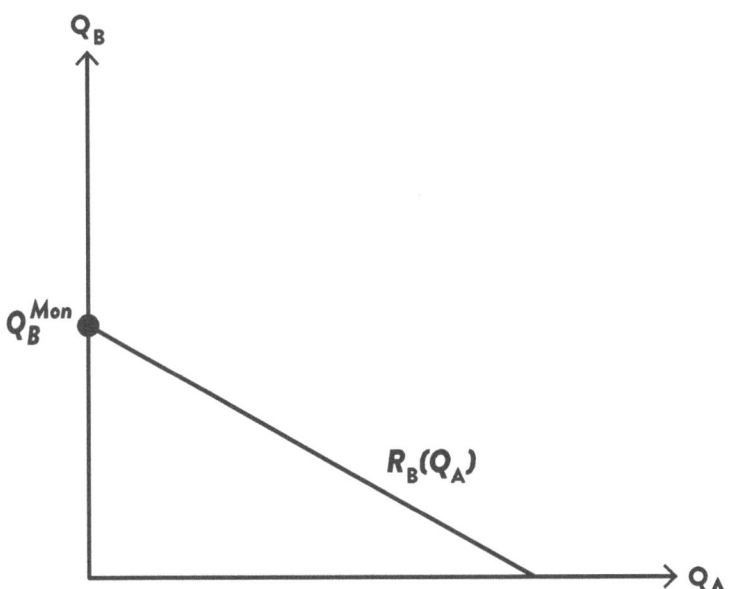

Fig. 12.2 Reaction function. *Note* The reaction function $R_B(Q_A)$ shows firm B's best response to different quantities produced by its competitor, firm A. If firm A produces nothing, firm B is a monopolist and chooses the monopoly quantity Q_B^{Mon}. When the competitor increases its quantity, firm B's best response is to reduce its own quantity. Therefore, the reaction function is downward sloping. For a sufficiently large quantity produced by firm A, firm B's best response is to produce nothing at all. This is indicated by the point where the reaction function crosses the horizontal axis

Conrad wonders what this analysis implies about how much production capacity they should install in their cardboard business. Brian explains that to answer this, they first need to determine what the market equilibrium will be. To do this, they must understand how their competitor thinks.

Brian: We can assume that they have exactly the same discussion as we do. We can also assume they have the same marginal costs as us; there's no reason to think otherwise, so the picture will look very similar.

Right now, Firstpak might be drawing a similar reaction function to the one I just showed, but for firm A: $R_A(Q_B)$. When we put these two reaction functions together, we get the picture shown in Fig. 12.3.

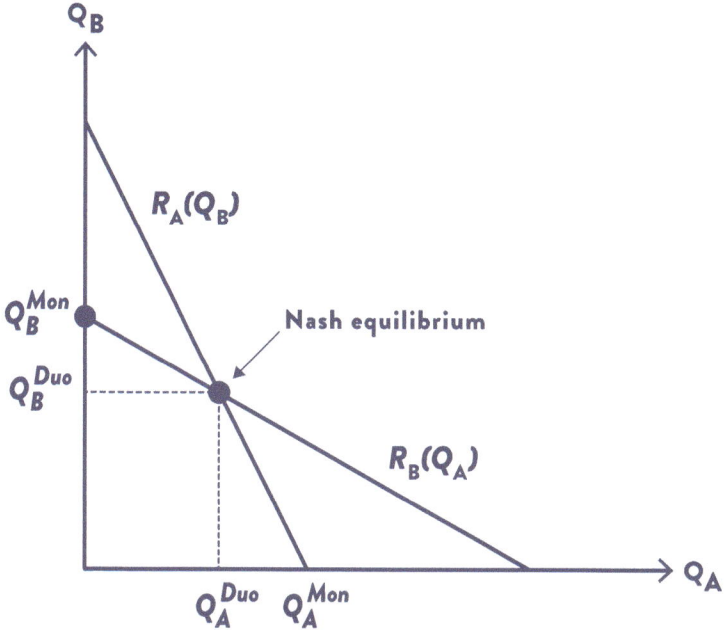

Fig. 12.3 Nash equilibrium. *Note* The figure shows the Nash equilibrium at the point where the two reaction functions intersect. At this point, neither firm has an incentive to change its choice, given the other's choice, and this is what defines a Nash equilibrium

My guess is that we'll end up producing quantities given by the point where the reaction functions intersect, so we should set our production accordingly, as shown in the figure by Q_A^{Duo} and Q_B^{Duo}.

Conrad: You'll have to slow down a bit, Brian. Why do you think we end up with exactly that quantity?

Brian: Because that's the only situation where neither we nor Firstpak has any reason to change our choices. Imagine, for example, that we chose quantity Q_B^{Mon}—we know that's our best choice if we were alone in the market. But will we be alone? No. Looking at Firstpak's reaction function, $R_A(Q_B)$, it's clear they won't let us be alone we can see from their reaction function what their best response to Q_B^{Mon} would be, and it's a positive quantity. And then Q_B^{Mon} isn't a smart choice for us.

Conrad: I see. So, this keeps going back and forth. I think I understand now. The only point where both firms are happy with their choices is when they produce Q_A^{Duo} and Q_B^{Duo}.

Brian: Exactly, Conrad. At the School of Economics, we called this a Nash equilibrium.

Nash equilibrium is named after a mathematical genius who struggled with schizophrenia for much of his adult life: John Nash (1928–2015). He was awarded the Nobel Prize in Economics in 1994, largely for his doctoral thesis, which he completed at the young age of 22.

Nash is portrayed in the book and film of the same name, *A Beautiful Mind*, which some of you may have read or seen (if not, you should!).

"I can work it out if you want." Without waiting for a reply, Brian pulls out paper and pencil and derives the Nash equilibrium for a Cournot duopoly, as shown in Math Box 12.1.

Math Box 12.1: Nash-Cournot Equilibrium

The profit of firm A is:

$$\pi_A = P(Q)Q_A - C(Q_A) \quad \text{Profit firm A}$$

And that of firm B is:

$$\pi_B = P(Q)Q_B - C(Q_B) \quad \text{Profit firm B}$$

Assume that demand in inverse form is given by:

$$P = 1 - Q \quad \text{Inverse demand}$$

Assume constant marginal costs, c_A and c_B. The profit functions of the two firms can then be written as:

$$\pi_A = (1 - Q_A - Q_B)Q_A - c_A Q_A$$

$$\pi_B = (1 - Q_A - Q_B)Q_B - c_B Q_B$$

Firm A maximises its profit with respect to its quantity Q_A, which is the strategic variable. The first-order condition is then given by:

$$\frac{\partial \pi_A}{\partial Q_A} = 1 - 2Q_A - Q_B - c_A = 0$$

From this expression, we can find firm A's reaction function as:

$$Q_A = \frac{1 - Q_B - c_A}{2} = R_A(Q_B) \quad \text{Firm A's reaction function}$$

This expression shows the quantity that firm A will choose for any quantity chosen by its competitor, firm B. Since the two firms are completely symmetrical, the reaction function for firm B will be:

$$Q_B = \frac{1 - Q_A - c_B}{2} = R_B(Q_A) \quad \text{Firm B's reaction function}$$

The Cournot equilibrium is found where the reaction functions intersect. We can find this point by substituting Q_B from firm B's reaction function into firm A's reaction function. We begin by expressing firm A's reaction function as follows:

$$Q_A = \frac{1 - c_A}{2} - \frac{1}{2}Q_B$$

Inserting $R_B(Q_A)$ for Q_B we get:

$$Q_A = \frac{1 - c_A}{2} - \frac{1}{2}\left(\frac{1 - Q_A - c_B}{2}\right)$$

$$Q_A - \frac{1}{4}Q_A = \frac{1 - c_A}{2} - \frac{1 - c_B}{4}$$

$$\frac{3}{4}Q_A = \frac{1}{4}(1 - 2c_A + c_B)$$

In Cournot equilibrium, firm A therefore produces:

$$Q_A^{Duo} = \frac{1}{3}(1 - 2c_A + c_B) \quad \text{Firm A's production in Cournot equilibrium}$$

And because of symmetry, firm B's production must be:

$$Q_B^{Duo} = \frac{1}{3}(1 - 2c_B + c_A) \quad \text{Firm B's production in Cournot equilibrium}$$

Total production then becomes:

$$Q_{Duo} = Q_A^{Duo} + Q_B^{Duo} = \frac{1}{3}(2 - c_A - c_B)$$

And the resulting price:

$$P_{Duo} = 1 - Q_{Duo} = \frac{1}{3}(1 + c_A + c_B)$$

Note that if the two firms have identical marginal costs, $c_A = c_B = c$, then the expressions for quantity and price in equilibrium simplify to:

$$Q_{Duo} = \frac{2}{3}(1 - c)$$

$$P_{Duo} = \frac{1}{3}(1 + 2c)$$

12.4   The Stackelberg Model

So far, we have looked at a situation where firms make decisions simultaneously. But now imagine that Firstpak can get ahead of Conrad's Box and be the first to install new production capacity. This is not an unreasonable scenario in a situation where Firstpak has already been in the market for some time and is replacing its machinery, while Conrad is the newcomer.

We are then in a situation of Stackelberg competition, with Firstpak as the leader and the Box as the follower. How will Firstpak deal with this head start? They begin by thinking backwards! That is, they ask themselves how Conrad will respond if they show off their brand-new cardboard machines. We call this backward induction. Firstpak analyses how Conrad will respond and then chooses their quantity based on this.

Look at Fig. 12.4 with Firstpak as firm A and the Box as firm B. In short, what Firstpak does is choose its favourite point on the competitor's reaction function, $R_B(Q_A)$. In the figure, this is b, where Firstpak chooses Q_A^{Lead} and Conrad responds by choosing Q_B^{Follow}.

How can we know that Firstpak, as leader, will increase its quantity compared to the Cournot equilibrium at point a? This is not obvious from the figure, but we can calculate it! See Math Box 12.2.

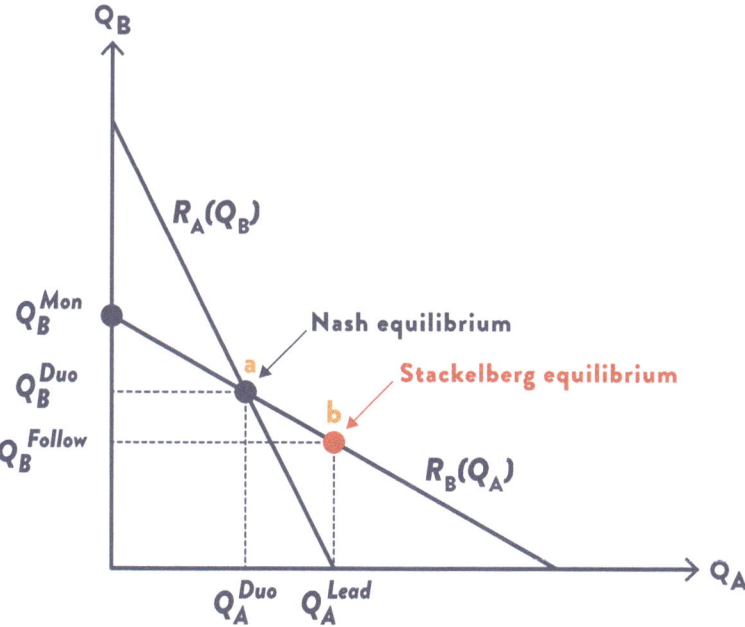

Fig. 12.4 Stackelberg equilibrium. *Note* The leader, firm A, chooses a larger quantity than in the Nash equilibrium, and the follower, firm B, responds by choosing a point on its reaction function with lower production than in the Nash equilibrium. In this way, the leader obtains a dominant market position and higher profit

Conrad in a heated discussion with the competitor at Firstpak

Math Box 12.2: Stackelberg Equilibrium

We start from the same demand and cost structure as in Math Box 12.1, with demand $P = 1 - Q$ and constant marginal costs c_A and c_B.

Backward induction means that the leader asks: How will the competitor respond to my choice of production capacity?

The answer to this question we have already found in Math Box 12.1: It is the follower (firm B)'s reaction function:

$$Q_B = \frac{1 - Q_A - c_B}{2} = R_B(Q_A) \quad \text{Firm B's reaction function}$$

The leader, firm A, takes the follower's response into account when choosing its quantity. Firm A therefore uses $Q_B = R_B(Q_A)$ in its profit expression:

$$\pi_A^{Lead} = \left(1 - Q_A - \left(\frac{1 - Q_A - c_B}{2}\right)\right)Q_A - c_A Q_A$$

This can be written as:

$$\pi_A^{Lead} = \frac{1}{2}(1 - Q_A + c_B)Q_A - c_A Q_A \quad \text{Firm A (leader) profit function}$$

The leader maximises profit to find the optimal quantity, taking the follower's response into account:

$$\frac{\partial \pi_A^{Lead}}{\partial Q_A} = \frac{1}{2}(1 - 2Q_A + c_B) - c_A = 0 \quad \text{Profit maximisation leader}$$

From this we can find the leader's optimal quantity as:

$$Q_A^{Lead} = \frac{1}{2}(1 - 2c_A + c_B) \quad \text{Quantity Stackelberg leader}$$

The follower's choice is then found by plugging the leader's quantity into the follower's reaction function:

$$Q_B^{Follow} = \frac{1 - Q_A^{Lead} - c_B}{2}$$

$$Q_B^{Follow} = \frac{1 - \left(\frac{1}{2}(1 - 2c_A + c_B)\right) - c_B}{2}$$

$$Q_B^{Follow} = \frac{1}{4}(1 - 3c_B + 2c_A) \quad \text{Quantity Stackelber follower}$$

Total production in Stackelberg-equilibrium is then:

$$Q_{Sta} = Q_A^{Lead} + Q_B^{Follow} = \frac{1}{4}(3 - 2c_A - c_B) \quad \text{Total quantity Stackelberg equilibrium}$$

And the Stackelberg-equilibrium price is:

$$P_{Sta} = 1 - Q_{Sta} = \frac{1}{4}(1 + 2c_A + c_B) \quad \text{Price Stackelberg equilibrium}$$

Note that if the firms have identical marginal costs, $c_A = c_B = c$, then the leader's production in equilibrium becomes:

$$Q_A^{Lead} = \frac{1}{2}(1 - c)$$

And that of the follower:

$$Q_B^{Follow} = \frac{1}{4}(1 - c)$$

We see that the leader here produces twice as much as the follower. This is illustrated in Fig. 12.4, where the leader A increases its production compared to the Cournot duopoly, thereby forcing the follower B to choose a lower quantity.

The total production in the Stackelberg equilibrium with identical marginal costs becomes:

$$Q_{Sta} = \frac{3}{4}(1 - c)$$

And the price:

$$P_{Sta} = 1 - Q_{Sta} = \frac{1}{4}(1 + 3c)$$

If we compare Stackelberg and Cournot, as found in Math Box 12.1, and for simplicity focus on the situation where the firms have identical marginal costs, we see that:

$$Q_{Sta} = \frac{3}{4}(1 - c) > Q_{Duo} = \frac{2}{3}(1 - c)$$

The leader–follower model thus results in a higher total quantity produced than the Cournot model, and the price will accordingly be lower. This is illustrated in Fig. 12.5, which also includes the market outcomes under monopoly and perfect competition to put the duopoly outcomes into context.

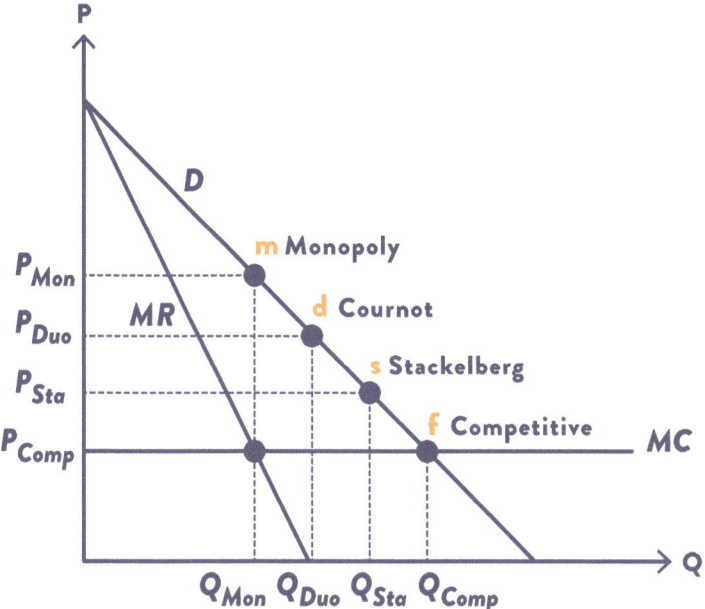

Fig. 12.5 Comparing market outcomes. *Note* The figure illustrates market outcomes ranging from monopoly to perfect competition. We see that a monopoly leads to the highest price, while perfect competition results in the lowest. A market with Cournot competition yields a higher price than one with Stackelberg competition

12.5 The Bertrand Model

As you've probably gathered by now, Conrad has a lot going on. And there's no time to rest: a competing publishing house, Beta Books, has published a new textbook in microeconomics, thereby challenging Alpha Book's ABC.

Conrad pays a visit to Anna, who is responsible for sales at the publishing house, to discuss the new situation. Unlike the production of cardboard, which Conrad handles in-house, Alpha Books does not print its own books. Instead, they purchase these services from various printing houses across the country, and the books are delivered at marginal cost. There is more than enough capacity in the printing industry, so the marginal cost is constant regardless of how many books Conrad decides to order.

With no binding capacity constraints, we are now in a situation of price competition between the two publishers. We will begin by examining the Bertrand model with differentiated products, before moving on to the case where the firms produce a homogeneous good—which gives rise to the so-called Bertrand paradox. But more on that later.

The two microeconomics textbooks cover the same topics but in different ways. Some prefer the ABC because it is fun to read and contains some nice illustrations, while others favour the somewhat drier book published by Beta Books, which gets straight to the point without any distractions. In other words, the books are differentiated products.

Demand for the ABC depends on both its own price and the price of the competing product: a higher own price means lower demand, while a higher price on the competing product increases demand for the ABC as it makes it relatively cheaper.

Conrad is wondering what price they should set for their microeconomics book, now that the competing book is in the market. Anna starts with a calculation, as shown in Math Box 12.3.

Math Box 12.3: Bertrand-Competition with Differentiated Products

Assume that the demand for the good produced by firm A—which we can think of as the ABC—can be written as:

$$Q_A = 1 - 2p_A + p_B \quad \text{Demand for good A}$$

Similarly, the demand for the good produced by firm B can be written as:

$$Q_B = 1 - 2p_B + p_A \quad \text{Demand for good B}$$

The marginal cost for producer A is c_A and for producer B c_B, meaning both firms face constant marginal costs. The profit for firm A can then be written as:

$$\pi_A = p_A Q_A - c_A Q_A = (p_A - c_A)Q_A = (p_A - c_A)(1 - 2p_A + p_B) \quad \text{Profit firm A}$$

And for B:

$$\pi_B = p_B Q_B - c_B Q_B = (p_B - c_B)Q_B = (p_B - c_B)(1 - 2p_B + p_A) \quad \text{Profit firm B}$$

Maximising the profit of firm A with respect to the price of its good, we find the optimum as:

$$\frac{\partial \pi_A}{\partial p_A} = 1 - 4p_A + p_B + 2c_A = 0 \quad \text{Profit maximisation for firm A}$$

This can be expressed as:

$$p_A = \frac{1 + p_B + 2c_A}{4} = R_A(p_B) \quad \text{Reaction function firm A}$$

This is firm A's reaction function, $R_A(p_B)$. It shows firm A's optimal response to a price p_B set by its competitor. Due to symmetry, we can easily find firm B's reaction function as:

$$p_B = \frac{1 + p_A + 2c_B}{4} = R_B(p_A) \quad \text{Reaction function firm B}$$

The Bertrand equilibrium is found where the reaction functions intersect.

To simplify, we assume identical marginal costs, $c_A = c_B = c$, and substitute p_B from firm B's reaction function into firm A's reaction function:

$$p_A = \frac{1 + \left(\frac{1 + p_A + 2c}{4}\right) + 2c}{4}$$

$$4p_A - \frac{1}{4}p_A = \frac{5}{4} + \frac{5}{2}c$$

$$\frac{15}{4}p_A = \frac{5}{4}(1 + 2c)$$

$$P_A^{Duo} = \frac{1}{3}(1 + 2c) \quad \text{Firm A's price in Bertrand equilibrium}$$

And due to symmetry:

$$P_B^{Duo} = \frac{1}{3}(1 + 2c) \quad \text{Firm B's price in Bertrand equilibrium}$$

From Math Box 12.3, we notice that the firms' reaction functions in Bertrand competition are upward sloping, as illustrated in Fig. 12.6. If one firm raises its price, the best response for the competitor is to do the same. This is known as *strategic complements*, where choices move in the same direction.

Under Bertrand competition, the firms' profits increase along their reaction functions. For example, if firm A raises the price of its product, it will lose some customers to firm B, which in turn will experience higher sales without doing anything. However, the optimal response for B to A's higher price is to raise its own price, allowing B to earn even greater profits. The same reasoning applies to firm A if B chooses to increase its price. Thus, profits for both firms increase as they move outward along their respective reaction functions in the diagram.

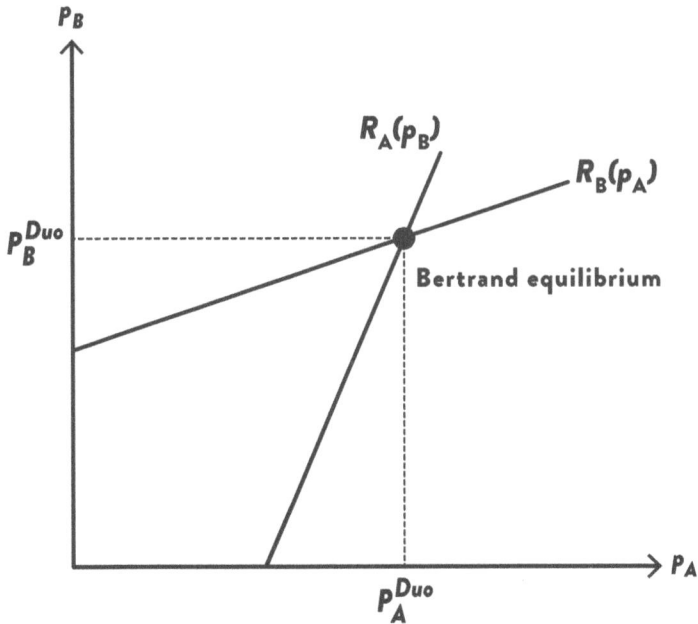

Fig. 12.6 Bertrand equilibrium, differentiated products. *Note* The figure shows the Bertrand equilibrium as the point where the two reaction functions intersect. Notice that the reaction functions are upward sloping in Bertrand competition. This means prices are strategic complements: if one firm raises its price, the best response for the competitor is to raise theirs as well

In the example of price competition between the publishers, we assumed that the two goods were differentiated products, meaning there is some difference between the two textbooks. But what if the goods are perfect substitutes—essentially identical. In such a situation, it is easy to see that the Nash equilibrium in price competition must be that the producers set prices equal to marginal cost, the same outcome as in perfect competition.

This is, in many ways, paradoxical since there are only two firms involved. In the literature, this is known as the *Bertrand paradox.*

Why does price equal marginal cost in equilibrium? Imagine a situation where both firms, A and B, initially charge a monopoly price. As we know, this maximises their joint profit. But this clearly cannot be an equilibrium because firm A can capture the entire market by setting its price slightly below its competitor's price. Firm B responds by lowering its price marginally below that of its competitor, and this price-cutting continues until there is no room for further reductions without incurring losses—which means the price equals marginal cost. In other words, the outcome is the same as in perfect competition.

An important reason why markets with two—or a few—producers do not end up in such price wars is that the goods they produce are not homogeneous but differentiated. Another key reason is limited production capacity, where the temptation to cut prices is restrained by the fact that a firm cannot serve the entire market demand anyway. Such markets can better be described by the Cournot model, where the price ends up somewhere between perfect competition and monopoly, and producers earn positive profits.

Yet another reason for the absence of price wars could be that firms cooperate to keep prices high, but this is illegal, and a well-functioning Competition Authority will put a stop to such behaviour.

12.6 Summary

In this chapter, we have examined the market structure known as oligopoly, where a small number of producers compete in the market. We explored three main models: quantity competition where producers choose simultaneously, also called Cournot competition; quantity competition where choices are made sequentially, known as the leader–follower model or Stackelberg competition; and price competition, known as Bertrand competition.

The equilibrium in these markets is referred to as a Nash equilibrium, defined as a situation in which no individual producer has an incentive to regret their choice.

We explained that quantity competition is most relevant in industries with binding capacity constraints, using the market for cardboard as an example. Price competition occurs in markets with unlimited capacity, illustrated by the publishing industry example, where publishers can outsource printing to various printing houses that collectively have ample capacity.

We also showed that simultaneous choices can be interpreted as producers not knowing each other's decisions, while sequential choices represent a structure where one firm moves before the other. In Stackelberg competition, this means the leader sets quantity before the follower. This gives the leader an advantage: by choosing a large quantity, the leader forces the follower to reduce theirs, thereby securing a dominant market position and higher profits.

12.7 Key Terms

Cournot Competition: An oligopoly model with quantity as the strategic variable, where firms choose quantity simultaneously.

Stackelberg Competition: An oligopoly model with quantity as the strategic variable, where firms choose quantity sequentially.

Bertrand Competition: An oligopoly model with price as the strategic variable, where firms choose price simultaneously.

Simultaneous Moves: A situation in which the players make their choices at the same time, meaning they do not know each other's choices.

Sequential Moves: A situation in which the players make their choices sequentially, meaning the follower observes the leader's decision.

Residual Demand: The demand that remains after the competitor has chosen its quantity.

Strategic Substitutes: When an increase in the strategic variable by one firm leads to a decrease in the strategic variable by the other.

Strategic Complements: When an increase in the strategic variable by one firm leads to an increase in the strategic variable by the other.

Reaction Function: Shows the best response to the competitor's choice.

Nash Equilibrium: A situation in which no player has a reason to regret their choice.

Bertrand Paradox: When two firms produce a homogeneous product, price competition will lead to an equilibrium with price equal to marginal cost.

12.8   Multiple-Choice Exercises

12.1: Residual Demand
Suppose there are three producers in the market. The inverse demand curve is given by $P = 10 - Q$, where $Q = Q_A + Q_B + Q_C$.

If firm A expects firms B and C to each produce 2.5 units, what residual demand will firm A face?

A. $P = 2.5 - 2Q_A$
B. $P = 5 - Q_A$
C. $P = 7.5 - Q_A$
D. $P = 5 - 2Q_A$

12.2: Cournot Equilibrium
A Cournot equilibrium is a situation where:

A. The producers maximise joint profits
B. None of the producers regret their choice of price
C. None of the producers regret their choice of quantity
D. None of the producers earn pure profits

12.3: Leader–Follower Model
Which of the following statements is correct when comparing a Stackelberg equilibrium with a Cournot equilibrium?

A. The leader wins, the follower loses, consumers win
B. The leader wins, the follower loses, consumers lose
C. Both the leader and the follower win, consumers lose
D. Both the leader and the follower lose, consumers win

12.4: Strategic Complements

What does it mean that prices in Bertrand competition are strategic complements?

A. If one firm increases its price, the other firm will increase its profit, but only if the two firms produce complementary goods
B. If one firm increases its price, the other firm's best response is to increase its own price
C. If one firm increases its price, the other firm's best response is to lower its own price
D. If one firm increases its price, the other will increase its quantity if the two firms produce complementary goods

Solutions: 12.1 B; 12.2 C; 12.3 A; 12.4 B

Game Theory

13

Why the cardboard producers are not able to maximise joint profits.

And why Anna's threat to punish Brian for his lack of effort during the first year of studies was not credible.

13.1   Introduction

Conrad thinks it is unfortunate that the competition between his own publishing house Alpha Books and Beta Books is so fierce. He would have preferred a gentleman's agreement, with both sides refraining from undercutting each other's prices. At the same time, he knows that competition law prohibits fixing prices. But what if Conrad on his own chooses to set a high price? Will Beta Books then respond in kind, so that they achieve peaceful coexistence without exchanging a word?

"Iacta alea est," exclaims Conrad at a crucial stage of a round of Ludo with his competitor. By the way, did you know that ludo means "I play" in Latin?

In the previous chapter, we studied competition between two firms that choose quantity or price to maximise their profits. But firms make many other strategic choices besides quantity and price, such as marketing, innovation, timing of new

K. Bjorvatn, *Microeconomics Made Simple*, Classroom Companion: Economics, https://doi.org/10.1007/978-3-032-06354-0_13

product launches, etc. Game theory is a flexible tool that makes it possible to analyse all kinds of strategic choices firms face in oligopoly markets.

And it's not only firms that think strategically! Game theory can be applied in many areas of life marked by tension and conflict—yes, everything from war to love. Take, for example, your everyday life as a student. You are going to work together in a group and give a presentation in microeconomics, and someone might think: "If I keep a low profile, the others in the group will do most of the work, so I choose to keep a low profile." This is a game between you and the others in the group.

In this chapter, we will study the most well-known games in the literature, with the colourful names Prisoner's Dilemma, Stag Hunt, and Chicken. I will also introduce you to the two most fundamental methods in game theory: normal form games, which we use when players move simultaneously, and extensive form games, when players move sequentially.

You have already seen examples of simultaneous and sequential games in the previous chapter: the Cournot and Bertrand models are games with simultaneous moves, while the Stackelberg model is a game with sequential moves. In these games, the strategic variable is continuous, whereas in game theory we typically consider a limited number of choices, for example, high or low price.

We begin with simultaneous moves and look at the most famous game of them all: Prisoner's Dilemma. We describe the Nash equilibrium in this game, and then move on to games with multiple equilibria, where the theory therefore does not give clear answers on what the outcome will be. Next, we study games with sequential moves, and how we use backward induction (that is, starting with the choice of the player who moves last) to find the equilibrium.

What can you do to ensure that the others in the group don't leave all the work to you? Perhaps you will be able to devise a credible threat after reading this chapter?

13.2 Prisoner's Dilemma

We have seen in the discussion about oligopoly that the total profit for the firms is lower than in a monopoly. It would be better for the duopoly firms if they did not compete so aggressively, for instance by agreeing to set a high price on their books. While such price fixing is prohibited under competition law, as long as the firms do not communicate, they can choose whichever price they want. So how about if they, without exchaning a word, set a high price?

Table 13.1 shows the game between Conrad's Alpha Books and the competing Beta Books as a matrix, also called a normal form game, which is used to analyse simultaneous games.

The numbers in each cell show the profits for the two producers, the first for the row player (here: Alpha Books) and the second for the column player (here: Beta Books). For example, look at the situation "High price, High price," where the

Table 13.1 Normal form game for Simultaneous Moves. *Note* The table shows the profits of the two players for different combinations of High and Low price

		BETA BOOKS	
		Low price	High price
ALPHA BOOKS	Low price	1, 1	3, 0
	High price	0, 3	2, 2

first term refers to the row player's choice and the second to the column player's choice. Here, the profit for each of them is 2 (million euros).

If Alpha Books chooses a low price and the competitor chooses a high price, Alpha Books will get a profit of 3 and Beta Books 0. We can think of this as Conrad capturing customers from his competitor by setting a lower price.

Looking at the profits in the different cells, the best outcome for both firms combined is to choose a high price. Together, they then have a total profit of 4. If both set a low price, the total profit is only 2. So, there are good reasons to stick to the high price, one would think. At the same time, it is tempting to undercut the competitor's price and thereby steal customers from him.

But what will be the outcome of this game? Let's tick off the best choices of each player, starting with Alpha Books. Conrad asks himself: If Beta Books has chosen a low price, what is my best option? Conrad looks at the low-price column and sees that by choosing a low price he earns 1, while a high price gives him 0. The choice is easy—he ticks off Low price, as shown in Table 13.2. The next question Conrad asks himself is: What if Beta Books has chosen a high price? Conrad looks at the high price column and sees that if he chooses a low price, he earns 3, but if he also sets a high price, he earns 2. Again, the answer is simple: tick off Low price.

Now let's look at the game from the perspective of Beta Books. If Alpha Books has chosen a low price, we see from the low-price row (looking at the second number) that Beta Books by choosing to also set a low price would earn 1, while a high price would give them 0. Beta Books therefore ticks off Low price.

Finally, what if Alpha Books sets a high price? Beta Books looks at the high-price row and sees that choosing a low price would give them 3 and a high price only 2, so they tick off Low price. The Nash equilibrium is the cell with two ticks, which here means that both firms choose a low price. Neither player has any reason to change their choice here.

The game described above is an example of the Prisoner's Dilemma, a situation where the players fail to achieve what is jointly best for them, which here would be for them to set a high price. The game was developed in the 1950s and describes a situation where two members of a criminal gang have been arrested, suspected

Table 13.2 Prisoner's Dilemma. *Note* The ticks mark each player's best choice, given the other player's choice. The Nash-equilibrium is the cell that both players have ticked off, in this case the Low price

<table>
<tr><td></td><td></td><td colspan="2" align="center">**BETA BOOKS**</td></tr>
<tr><td></td><td></td><td align="center">Low price</td><td align="center">High price</td></tr>
<tr><td rowspan="2">**ALPHA BOOKS**</td><td>Low price</td><td align="center">√1, 1√</td><td align="center">√3, 0</td></tr>
<tr><td>High price</td><td align="center">0, 3√</td><td align="center">2, 2</td></tr>
</table>

of various offenses. The dilemma for the prisoners is whether to betray the other and thereby receive a reduced sentence, or to stay silent. Both prisoners end up betraying the other, which places them behind bars for a long time.

From the game above, we see that Conrad chooses the low-price strategy regardless of what the competitor does. The same applies to Beta Books. Low price is thus the *dominant* strategy for both players. Similarly, setting a high price can be said to be a *dominated* strategy: the players will never choose it.

13.3   The Stag Hunt

Anna and Brian are sitting at their favourite café discussing climate change. Anna is an environmental activist, while Brian is concerned about the costs of a green transition. They agree that the world faces some difficult choices and decide to use game theory to study the issue. They divide the world into two parts, West and East, where the two regions can choose between Brown and Green technology. They imagine that the welfare from the different combinations of technology can be illustrated by the matrix in Table 13.3.

To solve this game, our friends use the tick-off method they learned in the microeconomics course at the School of Economics. If East chooses Brown technology, we see that West would do the same: West places a tick in the upper-left cell. If East chooses Green, West will also choose Green: West places a tick in the lower right cell. Since the situation is completely symmetrical for East, there will be two cells with two ticks: "Brown, Brown" and "Green, Green." This is therefore a situation with multiple equilibria. In addition to the two equilibria shown in the table, there is also an equilibrium in so-called mixed strategies, where the players flip a coin between the two strategies, but in this book, we focus on pure strategies, meaning that players do not randomise.

The most important difference between Stag Hunt and Prisoner's Dilemma is the values in the bottom-right cell: in Stag Hunt there is a lot to gain if both choose green technology—a big catch! And this is precisely the image that has given the game its name. Two wolves hunt hares or stag (an adult, male deer, in case you

Table 13.3 The Stag Hunt. *Note* The table shows the payoffs to the two players for different combinations of the strategies Brown and Green. The ticks mark each player's best choice, given the other's choice. There are two Nash equilibria in this case: both choose Brown or both choose Green.

were wondering). They can catch the hares on their own, but the big game requires cooperation. If both initially hunt hares, neither has an incentive to do otherwise, because catching a stag alone is nearly impossible. But if both hunt stag, this is also an equilibrium, since they realise this is the only realistic way to catch the big animal.

The parallel to the climate issue is clear. Research indicates there are critical threshold values for global warming: if the temperature rises above a certain limit, the ice in the Arctic and Greenland will start to melt, triggering major climate changes that may become self-reinforcing. To keep the temperature down, both East and West must make a green transition. Only then can they make the big catch, namely avoiding global warming.

One might think that such a scenario would make cooperation easy, but we know climate cooperation is difficult in practice. The Stag Hunt game in Table 13.3 shows why: when the starting point is a brown economy, neither West nor East has an incentive to initiate a green shift. This is a bad outcome, but it is an equilibrium.

This is thus a variant of the Prisoner's Dilemma, but the positive message is that there is also a good equilibrium and therefore hope! We may be hunting hares today, but tomorrow we can together hunt stag. We will soon study sequential games and see that if someone takes the lead in the transition to a green economy, they can get others on board and create a better world.

13.4   The Chicken Game

Alpha Books and Beta Books both have a new diet book in the works, and the question is when they should launch it. There are two good months for diet books: January (after Christmas dinners) and May (before bikini season), and of the two, January is slightly better since it is closer to New Year's resolutions.

Table 13.4 shows the profits for the publishers for the different combinations of release times, where the first number in each cell is the profit for Alpha Books and

Table 13.4 The Chicken Game. *Note* The table shows the payoffs to the two players for different combinations of strategies, which here are releasing the book in January or May. The ticks mark each player's best choice, given the other's choice. We see that both players mark "May, January" and "January, May," and there are therefore two Nash equilibria, where both players prefer to avoid doing the same.

BETA BOOKS

		January	May
ALPHA BOOKS	January	1.5, 1.5	$\sqrt{3}, 2\sqrt{}$
	May	$\sqrt{2}, 3\sqrt{}$	1, 1

the second is the profit for Beta Books. The worst scenario for the publishers is if both release the book in May, as competition is high and demand is not as strong as in winter. The dream scenario for each publisher is that their book comes out in January and the competing book in May.

The tick-off method shows that there are two Nash equilibria here: "January, May" and "May, January." This is yet another example of a game with multiple equilibria.

What is clear is that the outcome where both publishers release the diet book in the same month is not an equilibrium. But which book comes out after the Christmas dinners and which before the bikini season cannot be predicted by this game.

Games of the type described in Table 13.4 are known as Chicken. The name comes from a rather reckless competition where two drivers speed toward each other, and the one who swerves away first is a coward, that is, a "chicken." The worst outcome, however, is if neither of them swerves or if they turn and hit each other, and both die. The equilibrium is that one drives straight ahead and the other, the chicken, swerves away. But the game cannot predict who will do what. The point is that we have a situation where the two players most want to avoid doing the same. This also applies to the choice of release dates for the books, even though this is not a life-or-death game.

In the Chicken Game, there are no dominant or dominated strategies. Releasing the book in January can be best, but not always—it depends on the other player's choice. Likewise, releasing the book in May can be best, but not always. We note that the game predicts there will be a winner and a loser. The winner is the publisher who releases the book in January, while the loser must settle for May. The implication of this game thus contrasts with the Stag Hunt, where we saw that there is an equilibrium both players prefer.

13.5   Sequential Games

We remember from the chapter on oligopoly that sometimes it is natural to think of firms making decisions sequentially, meaning that one player moves before the other, like in chess. In chess, the white pieces make the first move, and this is often considered an advantage. This is also true in many of the games we will discuss here, but not always! Have you ever played "rock, paper, scissors"? That game is played simultaneously—one-two–three, reveal! But imagine a sequential version of this game: would you prefer to move first or last?

When analysing games with sequential moves, we use a slightly different tool called the extensive form game, often referred to as a game tree. And to solve this game, it's useful to cut branches! We start by setting up the Prisoner's Dilemma as a game tree and then look at the sequential variants of the Stag Hunt and the Chicken Game.

Sequential Prisoner's Dilemma

We used the pricing-strategy of the two publishing houses as an example of a Prisoner's Dilemma, as shown in Table 13.2. With simultaneous moves, both players chose a low price—the worst outcome for the two combined but the only equilibrium. Will things be different if, for example, Conrad moves first? Imagine that Conrad can commit to a high price by publishing his price list early on.

We draw the game as a tree, as shown in Fig 13.1, where four sub-branches are attached to two main branches. The meeting point between two branches is called a node. The node where the two main branches meet represents the first mover's choice, which here we think of as Alpha Books. Conrad must choose between a high and a low price. The two nodes attached to the four sub-branches represent the second mover's choices, those of Beta Books. They must also choose between a high or a low price, and the game thus has four end points. Below each end point are two numbers. The first represents the profit of the first mover, the second the profit of the second mover. You will see that the four pairs of numbers at the bottom of the game tree are the same as in the four cells in the normal form game in Tables 13.1 and 13.2.

In Fig. 13.2, we have brought the saw along to find the equilibrium in the tree model. Using backward induction we ask: What is the competitor's best choice if Alpha Books chooses a high price? Well, we see that Beta Books' best choice (focus on the second of the two numbers!) is then to choose a low price, since 3 is higher than 2. We can therefore use the saw on the sub-branch "High price, High price," as it will not be chosen.

What if Alpha Books chooses a low price? Then Beta Book's best response is also to choose a low price. Pick up the saw and cut off the sub-branch "Low price, High price."

With fewer branches on the tree, Conrad has a clearer picture of the game's course: If he chooses a high price, the competitor responds with a low price, and if he chooses a low price, the competitor will do the same. Conrad then compares the profit in "High price, Low price," which gives profit 0, with "Low price, Low

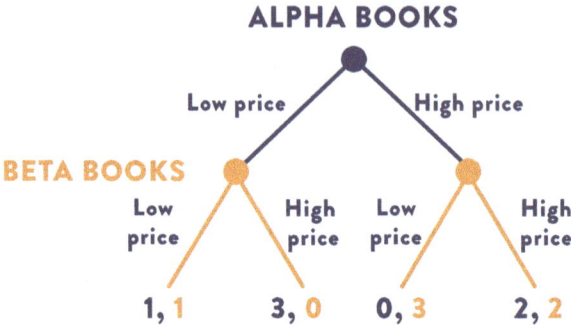

Fig. 13.1 Tree model for sequential moves. *Note* The figure shows the tree model for the sequential game where Alpha Books moves before Beta Books. The first number refers to the first mover's profit, the second to the second mover's profit. For example, if Alpha Books chooses a low price and Beta Books responds with a high price, Alpha Books gets 3 and Beta Books 0

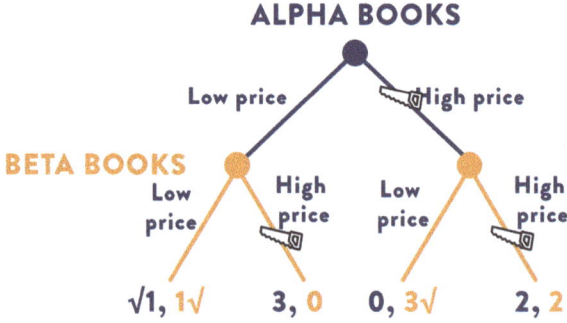

Fig. 13.2 Equilibrium in a sequential prisoner's dilemma. *Note* The figure shows the equilibrium in the tree model for the sequential game where Alpha Books moves before Beta Books. We use backward induction, which means we start with the second mover's best responses to the first mover's different choices. We tick off the second mover's best choices and cut off the branches that will not be used. We then tick off the best choice of the first mover, based on the remaining sub-branches, and cut off the main branch that will not be used. The equilibrium is the combination of choices that both players have ticked off. In this case, the equilibrium is that both players choose a low price

price," which gives profit 1, cuts off the former branch, and ticks off Low price. The Nash equilibrium, the outcome with two ticks and the only part of the tree still alive, is therefore that both firms choose the low price.

Strictly speaking, a Nash equilibrium in sequential games is called a subgame perfect Nash equilibrium. But for simplicity, we will here just refer to it as a Nash equilibrium.

We note that the sequential variant of this game gives the same equilibrium as the simultaneous one, namely low prices. Allowing one player to move first therefore makes no difference. But this is not always the case, as we will now see.

Sequential Stag Hunt

Anna and Brian are still sitting at the café discussing environmental issues. Anna argues that the West should take a leadership role and carry out a green transition (choose Green), but Brian argues that this would be too costly and cause a welfare loss for the West. Anna disagrees. Who is right?

Anna sets up the game shown in Table 13.3 as a tree, as shown in Fig. 13.3, with West as the first mover and East as the second mover. She saws off the branches "Brown, Green" and "Green, Brown". West's choice is now between "Brown, Brown" and "Green, Green," and the choice is simple: go for the green transition.

Brian was mistaken because he did not see the equilibrium in the game. West choosing Green is unprofitable if East does nothing (i.e., sticks with Brown technology). But how will East respond to West choosing Green? They will also choose a green transition, and this brings environmental benefits for West as well. So, Anna is right: It pays off for the West to take environmental leadership, and this creates a welfare gain for the whole world. Note that the sequential variant of the Stag Hunt gives a clear prediction of the outcome, whereas the simultaneous-move variant had two equilibria.

Sequential Chicken Game

The two publishers, Alpha Books and Beta Books, are thinking hard about when to release the new diet book: January or May. We remember that when the publishers moved simultaneously (see the simultaneous Chicken Game in Table 13.4), the game had two equilibria. But what if we now let one of the publishers, for example Alpha Books, move first? Alpha Books might have a more flexible organisation and a marketing department that can quickly advertise the new book to the market through coverage in the major newspapers.

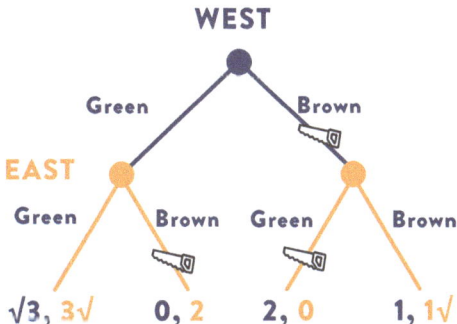

Fig. 13.3 Equilibrium in a sequential Stag Hunt. *Note* The figure shows the equilibrium in the tree model for the sequential Stag Hunt on the environment, where West moves before East. Using backward induction, we can cut the sub-branches "Brown, Green" and "Green, Brown," leaving the first mover with the choice between "Brown, Brown" and "Green, Green," and they choose the latter

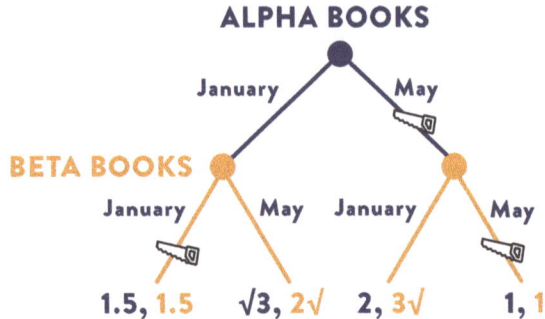

Fig. 13.4 Equilibrium in a sequential Chicken Game. *Note* The figure shows the tree model for the sequential game where Alpha Books can choose the release date of the book before Beta Books. The Nash equilibrium is that Alpha Books chooses January and Beta Books responds with May. The player who moves first receives a payoff of 3, while the one who moves last gets 2. There is thus a first-mover advantage in this game

The procedure is exactly as described above, and is shown in Fig. 13.4. We can cut the sub-branches "January, January" and "May, May," and the choice for Alpha Books is simple: they announce the release of the diet book in January, and Beta Books responds with May, so the Nash equilibrium is "January, May".

There are two interesting points to note about this game. First, we see that there is a difference between the simultaneous and sequential versions. In the simultaneous game, we found two equilibria, while in the sequential version, there is only one equilibrium. This is the same observation as for the Stag Hunt.

Second, we see that the player who moves first has an advantage: they will achieve a profit of 3, while the follower gets 2. This is therefore a game with a first-mover advantage. The sequential Chicken Game thus resembles the Stackelberg competition described in the previous chapter. There too, we found that the leader had an advantage by committing to a large quantity, thereby capturing market share from the follower.

13.6   A Credible Threat?

Anna and Brian are sitting at a café, looking back on their student days. As you probably remember, they studied together during their first year, before Brian moved to the School of Engineering. He says he thought there were too many mandatory assignments at the School of Economics, and besides, he was always more interested in mathematics and technology. Anna reflects upon the group work during that first year and how frustrated she was that she always ended up doing most of the work.

Anna and Brian at the café

Anna: Do you remember when I told you that if you didn't do your part, then I wouldn't do mine either, with the result being that we'd both fail? I ended up doing the whole job anyway…

Brian: I vaguely remember that, yes.

Anna: Yeah, it was a kind of game between us—and one that I lost.

Brian: Sorry.

Anna: By the way, do you remember the game theory lectures? The professor said you could use game theory to analyse all kinds of things. Shall we try to analyse the conflict about the group work?

Brian: I remember something about trees and branches that were supposed to be cut, but you know, I only half paid attention in class.

Anna: I thought game theory was one of the most fun parts of the micro course, and I remember we were supposed to use a game tree to analyse games with sequential structure. And since you were supposed to start the group assignment and I was supposed to finish it and hand it in, the group work was such a game. (Anna illustrates with Fig. 13.5).

Anna: If you, Brian, had been active, we would have passed without problems. We can assign a payoff of 2 to both of us for this outcome. If you didn't do your part, and I stayed passive, we would have failed—that gives zero payoff to both of us. But if you didn't do your part, and I did it for you, we would still pass, but it would cost me blood, sweat, and tears. You would pass without lifting a finger: payoff 3 for you, only 1 for me.

Anna's threat not to do anything if Brian didn't do anything was unfortunately not credible. In the figure, Brian sees that by choosing Passive, Anna faces the choice between being Passive herself, which gives her zero payoff, and being Active, which gives her payoff 1. She will then choose to be Active.

We can therefore disregard Anna's Passive strategy—it is an empty threat.

For Brian, the choice then stands between being Active himself or leaving the work to Anna. He chooses the latter.

BRIAN

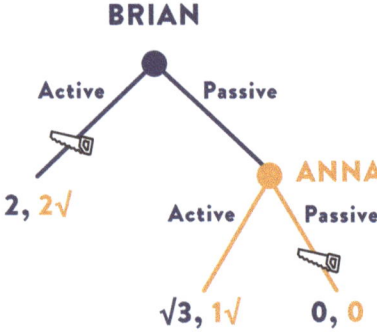

Fig. 13.5 Empty threat. *Note* The figure shows the game tree for the sequential game where Brian can choose to be active or passive in the group work with Anna. Anna will submit the assignment and therefore moves last. She threatens not to do anything if Brian also does nothing—that is, to choose Passive if Brian chooses Passive. But we see that this is not a credible threat. If Brian chooses Passive, Anna responds with Active. Brian prefers this outcome over being active himself, so "Passive, Active" becomes the Nash equilibrium in this game

Brian: Yes, that was an empty threat.

Anna: What should I have done?

Brian: I'm a bit ashamed of that period of my life. But I think you needed to do something that committed you not to do the work for me. You had a job in a grocery shop while studying, right? What if you had promised your boss that you could sit at the checkout all day before the deadline, and you told me this and even showed me the work schedule? If I didn't deliver and you had to do the work for me, you would also endure the stress of telling your boss you couldn't work that day after all. That would have changed your payoff for choosing Active, and the outcome could have been different. (Brian modifies the payoffs in the figure. He sets minus 1 as Anna's payoff for choosing Active if Brian has chosen Passive and shows Fig. 13.6. We see that Passive becomes a credible threat, and the outcome is now that Brian chooses to do his work.)

Anna: Yes, that sounds smart. I guess you did pick up something from the lectures after all!

A credible threat is about binding oneself to the mast. This expression comes from Homer's epic *The Odyssey*, where Odysseus, on his journey home from Troy, had to pass the sirens—female creatures who lured sailors with their enchanting song. To avoid being led into temptation, he stuffed wax in his crew's ears and asked them to tie him to the mast.

Fig. 13.6 Credible threat.
Note The figure shows the
sequential game where Anna
has reduced her payoff from
choosing Active if Brian
chooses Passive. In this case,
Anna responds to Brian's
Passive choice by choosing
Passive herself. This makes
Passive a credible threat from
Anna. Brian sees this and
chooses to be Active

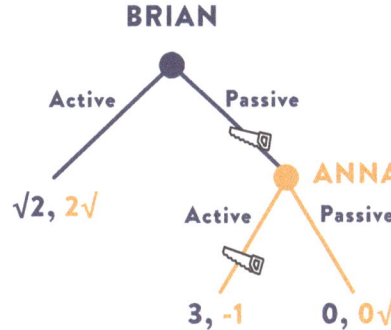

13.7   Summary

In this chapter, we have examined strategic interaction using game theory, a set
of tools that makes it possible to analyse many situations involving an element
of conflict. We have looked at simultaneous and sequential games, dominant and
dominated strategies, and unique and multiple equilibria.

Simultaneous games are typically presented in what is called the normal form
and we have learned the tick-off method to find the Nash equilibrium—or equilibria, in plural! Sequential games are presented in the extensive form, also referred
to as a game tree. We have learned to solve sequential games using backward
induction and seen that cutting branches can be a useful tool for finding the
equilibrium.

P.S. Anna and Brian later had several romantic meetings, both at the cinema and at cafés. They now live in a lovely house at a convenient distance from work, and have children, a golden retriever, and an electric bike with a trailer. Anna has taken over the factory from her grandfather, and Brian is responsible for making sure everything technical runs smoothly—in the paper mill, the publishing house, and the cardboard factory.

Conrad has retired and is writing a book about the history of the paper mill, from when it was founded by his father to today's extensive business with Anna and Brian at the helm. He looks at Anna and her growing family and notes with satisfaction that the final chapter of the family-owned company is yet to be written.

The final chapter of the family-owned business is yet to be written…

13.8   Key Terms

Prisoner's dilemma: A game where the equilibrium gives the lowest total payoff to the players.
Dominant strategy: A best choice regardless of what the opponent does.
Dominated strategy: A worse choice regardless of what the opponent does.
Stag hunt: A game with multiple equilibria where the players agree on which equilibrium is best.
Multiple equilibria: Games with more than one Nash equilibrium.
Chicken game: A game with multiple equilibria, where players disagree about which is best, but agree that the worst outcome is if they do the same.
First-mover advantage: a sequential game where it is advantageous to move first.
Empty threat: a threat of an action that will not actually be carried out.
Credible threat: a threat that will actually be carried out.

13.9   Multiple-Choice Exercises

13.1: Dominant and Dominated Strategies
Consider two players, Row and Column, who are playing a board game with the following choices and outcomes:

COLUMN

		Up	Down
ROW	**Up**	7, 1	0, 5
	Down	1, 4	4, 2

Which of the Following Statements is Correct?

A. Row's dominant strategy is Up
B. Row's dominated strategy is Up
C. Column's dominant strategy is Right
D. Column's dominated strategy is Right

13.2: Multiple or No Equilibria
Assume two players, Row and Column, who are playing a board game with the following choices and outcomes:

COLUMN

		Up	**Down**
ROW	**Up**	5, 6	0, 0
	Down	1, 3	4, 1

What is/are the game's Nash equilibrium(s)?

A. One equilibrium: (Up, Left)
B. One equilibrium: (Down, Right)
C. Multiple equilibria: (Up, Left) and (Down, Right)
D. There is no Nash equilibrium

13.3: Sequential Moves

The game described the figure below is exactly the same as the game in 13.2 regarding possible strategies and payoffs. The difference is that it is played sequentially, where Row chooses before Column.

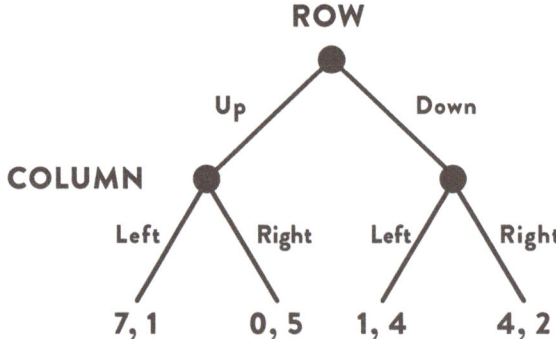

What is the Nash equilibrium in this sequential game?

A. Row chooses Up, Column chooses Left
B. Row chooses Up, Column chooses Right
C. Row chooses Down, Column chooses Left
D. Row chooses Down, Column chooses Right

13.4: Credible Threat

Consider the two games described the figure below between First and Second, where the choices are between Active (A) and Passive (P). In which game(s) can we say that Passive is a credible threat from the Second player's side?

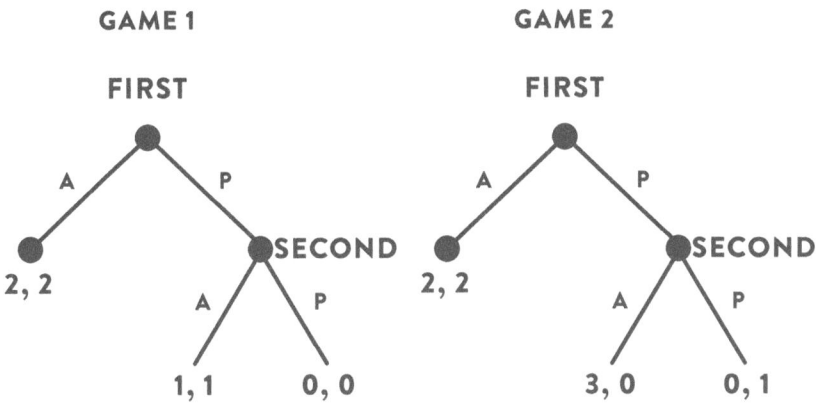

A. Game 1
B. Game 2
C. Both games
D. Neither game

Solutions: 13.1 D; 13.2 D; 13.3 C; 13.4 B.

Index

A
Alternative cost, 79, 109
Altruism, 67, 68, 70, 71
Average total cost (ATC), 108, 110–112
Average variable cost (AVC), 108, 110–112

B
Backward induction, 247–249
Behavioural economics, 60–62
Bertrand competition, 234
Bertrand equilibrium, 235
Bertrand, Joseph, 220
Bertrand model, 233
Bertrand paradox, 236
Budget constraint, 4
Budget line, 4–8

C
Capital, 76–80
Choke price, 143
Cobb-Douglas production function, 80
Cobb-Douglas utility function, 7
Completeness, 6, 7
Consumer, 4–13
Consumer price, 182–187
Consumer surplus (CS), 167
Consumption, 4–6
Cournot, Augustin, 220
Cournot competition, 220
Cournot model, 221
Cross price effect, 17–20

D
Deadweight loss, 166, 173–175
Decreasing returns to scale, 116

D
Dictator game, 67
Differentiated products, 236
Dominant strategy, 244, 255
Dominated strategy, 244, 255
Duopoly, 221

E
Economic costs, 109, 167, 180
Efficiency loss, 173
Elasticity of demand, 145–147
Equilibrium, 152–157
External effect, externality, 173

F
Factors of production, 80–83
First-mover advantage, 250
Fixed costs, 109
Fungibility, 34

G
Game in normal form, 242
Game theory, 241
Game tree, 247
Giffen good, 30
Giffen, Robert, 34

H
Hicks, John Richard, 168
Homo economicus, 59
Homogeneous product, 141

I
Income expansion path, 30

Import, 159
Income, 4–7
Income effect (IE), 26–28
Income target, 64–66
Indifference curve, 8–12
Inefficiency, 167
Input factor, 77
Internalise, 192
International trade, 158
Invisible hand, 166
Isocost line, 78–80
Isoquant, 81–87

K

Kahneman, Daniel, 60, 64
Kaldor-Hicks criterion, 168
Kaldor, Nicholas, 168

L

Labour, 77
Labour demand, 134
Labour supply, 46–48
Lagrange method, consumption, 12
Lagrange method, production, 86
Law of diminishing marginal product, 83
Leontief production function, 93, 112
Leontief, Wassily, 93
Long run, 16

M

Marginal cost (MC), 108
Marginal product of capital, 82–85
Marginal product of labour, 82–85
Marginal rate of substitution (MRS), 9
Marginal technical rate of substitution
        (MRTS), 82
Marginal utility, 9
Market equilibrium, 152
Market failure, 166–168, 173–175, 191
Market power, 204
Marshall, Alfred, 34
Matrix model, 244
Minimum wage, 180, 194
Multiple equilibria, 242, 244

N

Nash equilibrium, 225, 243
Nash, John, 226, 245
Natural monopoly, 215
Negative externality, 173

Normal goods, 29
Normative, 168

O

Oligopoly, 219
Operating profits (OP), 124–127
Opportunity cost, 47, 79, 109, 125

P

Perfect competition, 142
Pigou, Arthur Cecil, 192
Pigouvian tax, 191
Preference, 6–9
Price discrimination, 212
Price regulation, 180, 194, 216
Producer, 75
Producer price, 182–187
Producer surplus (PS), 166
Production, 75–78
Production function, 80
Profit, 124–136
Profit maximisation, 126

Q

Quantity competition, 220, 237
Quotas, 180, 194

R

Raw materials, 109, 181
Reaction function, 223–227
Real wage, 46
Residual demand (RD), 222
Ricardo, David, 168
Robots, 94–99

S

Scale properties, 120
Shadow price, 47, 79
Short run, 106
Simultaneous moves, 242
Smith, Adam, 166
Stackelberg competition, 228
Stackelberg model, 228
Strategic substitutes, 222
Subsidies, 180, 186, 193
Substitute goods, 19
Substitution effect (SE), 26–30
Sunk cost, 109
Supply curve, 132–135

T
Tariff, 189–191
Tax, 179–189
Thaler, Richard, 62
Time-inconsistent preferences, 61
Total surplus, 166
Two-period model, 36

U
Utility, 7–9
Utility function, 7–10

V
Variable costs (VC), 109–112
Von Stackelberg, Heinrich Freiherr, 220